Linear Algebra

Dr. K. Q. Lan

CUSTOM EDITION FOR RYERSON UNIVERSITY

Learning Solutions

New York Boston San Francisco
London Toronto Sydney Tokyo Singapore Madrid
Mexico City Munich Paris Cape Town Hong Kong Montreal

Cover Art: Courtesy of brandXpictures

Pearson Learning Solutions, 501 Boylston Street, Suite 900, Boston, MA 02116
A Pearson Education Company
www.pearsoned.com

Printed in Canada

3 4 5 6 7 8 9 10 XXXX 15 14 13 12 11 10

000200010270590884

BK

ISBN 10: 0-558-74636-5
ISBN 13: 978-0-558-74636-0

Contents

Chapter 1

Vectors and $\mathbb{R}^n$-space

1.1 Definition of vectors

We denote by $\mathbb{N}$ the set of positive integers, that is, $\mathbb{N} = \{1, 2, \cdots, \}$. Let $n \in \mathbb{N}$, where the symbol '$\in$' means 'belongs to', that is, n belongs to $\mathbb{N}$. Let $\mathcal{I}_n = \{1, 2, \cdots, n\}$ and let $\mathbb{R} = (-\infty, \infty)$ be the set of real numbers.

Definition 1.1.1. An ordered set of n numbers written as

$$(a_1, a_2, \cdots, a_n)$$

is called an n-row vector. If it is written as

$$\begin{pmatrix} a_1 \\ a_2 \\ \vdots \\ a_n \end{pmatrix},$$

then it is called an n-column vector.

We often refer to an n-row vector or an n-column vector as an n-vector or a vector. We denote a vector by $\overrightarrow{a}$, $\overrightarrow{b}$, $\overrightarrow{u}$, $\overrightarrow{v}$ and so on. a_i is called the ith component (ith entry) of the vector for each $i \in \mathcal{I}_n$.

Note that two vectors with the same components written in different orders may not be same. For example, the row vectors $\begin{pmatrix} 1 \\ 0 \end{pmatrix}$ and $\begin{pmatrix} 0 \\ 1 \end{pmatrix}$ are different vectors. Thus, the word "ordered" in the definition of a vector is essential.

Example 1.1.2. (1) $\overrightarrow{a} = (5)$ is a row (or column) vector (or 1-vector); $\overrightarrow{a} = (1, 2)$ is a row vector (or 2-vector); $\overrightarrow{a} = (1, 2, 0.5)$ is a row vector (or 3-vector) and $\overrightarrow{a} = (1, 2, \cdots, n)$ is a row vector (or n-vector).

(2) $\begin{pmatrix} 0 \\ 0 \\ 0 \end{pmatrix}$ is a column vector (or 3-vector) and $\begin{pmatrix} 1 \\ 2 \\ \vdots \\ n \end{pmatrix}$ is a column vector (or n-vector).

Definition 1.1.3. A vector whose components are all zero is called a zero vector . A vector which is not a zero vector is called a nonzero vector.

Note that if a vector has at least one nonzero component, it is a nonzero vector.

Example 1.1.4. $(0, 0, 0)$ is called a zero vector while $(1, 1)$, $(0, 2, 0)$ and $(1, 1, 0, 0)$ are nonzero vectors.

Exercises

1. Write a 3-column zero vector and a 5-column zero vector, respectively.
2. Write a 3-column nonzero vector and a 5-column nonzero vector, respectively.
3. Suppose that the buyer for a manufacturing plant must order different quantities of oil, paper, steel and plastics. He would order 40 units of oil, 50 units of paper, 80 units of steel and 20 units of plastics. Write the quantities in a single vector.
4. Suppose that a student's course marks for quiz 1, quiz 2, test 1, test 2 and final exam are 70, 85, 80, 75, 90, respectively. Write his marks as a column vector.

1.2 $\mathbb{R}^n$-space

Definition 1.2.1. The set of all n-column vectors is called $\mathbb{R}^n$-space .

We write the set as follows:

$$\mathbb{R}^n = \Big\{ \begin{pmatrix} a_1 \\ a_2 \\ \vdots \\ a_n \end{pmatrix} : a_i \in \mathbb{R} \quad \text{and } i \in \mathcal{I}_n \Big\}.$$

Example 1.2.2. $\mathbb{R}^2 = \left\{\begin{pmatrix} x \\ y \end{pmatrix} : x, y \in \mathbb{R}\right\}$, the set of all vectors in the xy-plane.

$\mathbb{R}^3 = \left\{\begin{pmatrix} x \\ y \\ z \end{pmatrix} : x, y, z \in \mathbb{R}\right\}$, the set of all vectors in the coordinate system xyz.

If $\overrightarrow{a} = \begin{pmatrix} a_1 \\ a_2 \\ \vdots \\ a_n \end{pmatrix}$ is a vector in the $\mathbb{R}^n$-space, we write $\overrightarrow{a} \in \mathbb{R}^n$.

Example 1.2.3. $\overrightarrow{a} = \begin{pmatrix} 1 \\ 2 \end{pmatrix} \in \mathbb{R}^2$ and $\overrightarrow{a} = \begin{pmatrix} 1 \\ 0 \\ 3 \end{pmatrix} \in \mathbb{R}^3$.

Let $\overrightarrow{e_1} = \begin{pmatrix} 1 \\ 0 \\ \vdots \\ 0 \end{pmatrix}, \overrightarrow{e_2} = \begin{pmatrix} 0 \\ 1 \\ \vdots \\ 0 \end{pmatrix}, \cdots, \overrightarrow{e_n} = \begin{pmatrix} 0 \\ 0 \\ \vdots \\ 1 \end{pmatrix}$, each has n components.
These vectors $\overrightarrow{e_1}$, $\overrightarrow{e_2}$, $\cdots$, $\overrightarrow{e_n}$ are called the standard vectors in $\mathbb{R}^n$.

Example 1.2.4. (1) Let $\overrightarrow{e_1} = \begin{pmatrix} 1 \\ 0 \end{pmatrix}$ and $\overrightarrow{e_2} = \begin{pmatrix} 0 \\ 1 \end{pmatrix}$. Then $\overrightarrow{e_1}$ and $\overrightarrow{e_2}$ are the standard vectors in $\mathbb{R}^2$.

(2) Let $\overrightarrow{e_1} = \begin{pmatrix} 1 \\ 0 \\ 0 \end{pmatrix}$, $\overrightarrow{e_2} = \begin{pmatrix} 0 \\ 1 \\ 0 \end{pmatrix}$ and $\overrightarrow{e_3} = \begin{pmatrix} 0 \\ 0 \\ 1 \end{pmatrix}$. Then $\overrightarrow{e_1}$, $\overrightarrow{e_2}$ and $\overrightarrow{e_3}$ are the standard vectors in $\mathbb{R}^3$.

Note that the set of all n-row vectors is also called $\mathbb{R}^n$-space. We often consider $\mathbb{R}^n$-space as the set of all n-column vectors and study properties of n-column vectors. Similar definitions and properties hold for n-row vectors.

Exercises

1. Determine in which $\mathbb{R}^n$-space are the following vectors.

$$\overrightarrow{a} = \begin{pmatrix} 1 \\ 2 \\ 3 \end{pmatrix} \quad \overrightarrow{b} = \begin{pmatrix} 0 \\ 2 \\ 3 \\ 0 \end{pmatrix} \quad \overrightarrow{c} = \begin{pmatrix} 1 \\ 0 \\ 3 \\ 3 \\ 3 \\ 7 \end{pmatrix} \quad \overrightarrow{d} = \begin{pmatrix} 0 \\ 2 \\ 0 \\ 0 \\ 1 \end{pmatrix}$$

1.3 Equality of two vectors

Definition 1.3.1. Two column vectors are said to be equal if the following two conditions are satisfied:

(i) The number of components of the two vectors must be same.

(ii) The corresponding components of the two vectors are equal.

If two vectors $\overrightarrow{a}$, $\overrightarrow{b}$ are equal, we write $\overrightarrow{a} = \overrightarrow{b}$. Otherwise, we write $\overrightarrow{a} \neq \overrightarrow{b}$.

Let $m, n \in \mathbb{N}$ and let $\overrightarrow{a} = \begin{pmatrix} a_1 \\ a_2 \\ \vdots \\ a_m \end{pmatrix}$ and $\overrightarrow{b} = \begin{pmatrix} b_1 \\ b_2 \\ \vdots \\ b_n \end{pmatrix}$. According to Definition 1.3.1, $\overrightarrow{a} = \overrightarrow{b}$ if and only if $m = n$ and $a_i = b_i$ for each $i \in \mathcal{I}_n$ (that is, $a_1 = b_1$, $a_2 = b_2$, $\cdots$, $a_n = b_n$). Moreover, if $\overrightarrow{a} \neq \overrightarrow{b}$, then one of (i) and (ii) in Definition 1.3.1 does not hold, that is, either $m \neq n$ or there exists some $i \in \mathcal{I}_n$ such that $a_i \neq b_i$.

Example 1.3.2. (1) Let $\overrightarrow{a} = \begin{pmatrix} x+y \\ x-y \\ z \end{pmatrix}$ and $\overrightarrow{b} = \begin{pmatrix} 1 \\ 3 \\ 1 \end{pmatrix}$. Find all $x, y, z \in \mathbb{R}$ such that $\overrightarrow{a} = \overrightarrow{b}$.

(2) Let $\overrightarrow{a} = \begin{pmatrix} 1 \\ x^2 \end{pmatrix}$ and $\overrightarrow{b} = \begin{pmatrix} 1 \\ 4 \end{pmatrix}$. Find all $x \in \mathbb{R}$ such that $\overrightarrow{a} \neq \overrightarrow{b}$.

(3) $\overrightarrow{a} = \begin{pmatrix} 1 \\ 2 \\ 3 \end{pmatrix}$ and $\overrightarrow{b} = \begin{pmatrix} 1 \\ 2 \end{pmatrix}$. Identify whether $\overrightarrow{a}$ and $\overrightarrow{b}$ are equal.

Solution. (1) Since the numbers of components of $\overrightarrow{a}$ and $\overrightarrow{b}$ are equal, by Definition 1.3.1, in order to make $\overrightarrow{a} = \overrightarrow{b}$, we need the corresponding components of $\overrightarrow{a}$ and $\overrightarrow{b}$ to be equal, that is, x, y, z satisfy the following system of equations:

$$\begin{cases} x+y=1 \\ x-y=3 \\ z=1 \end{cases}$$

Solving the above system, we get $x = 2$, $y = -1$ and $z = 1$. Hence, when $x = 2$, $y = -1$ and $z = 1$, $\overrightarrow{a} = \overrightarrow{b}$.

(2) Since the numbers of components of $\overrightarrow{a}$ and $\overrightarrow{b}$ are equal, by Definition 1.3.1, if $x^2 \neq 4$, then $\overrightarrow{a} \neq \overrightarrow{b}$. Solving $x^2 \neq 4$, we get $x \neq 2$ and $x \neq -2$. Hence, when $x \neq 2$ and $x \neq -2$, $\overrightarrow{a} \neq \overrightarrow{b}$.

(3) Since the numbers of components of $\overrightarrow{a}$ and $\overrightarrow{b}$ are 3 and 2, respectively, it follows from Definition 1.3.1 that $\overrightarrow{a} \neq \overrightarrow{b}$. □

We often treat a column vector $\overrightarrow{a} = \begin{pmatrix} a_1 \\ a_2 \\ \vdots \\ a_n \end{pmatrix}$ as a row vector $\overrightarrow{a} = (a_1, a_2, \cdots, a_n)$.

Hence, if $\overrightarrow{a} = (a_1, a_2, \cdots, a_m)$ and $\overrightarrow{b} = \begin{pmatrix} b_1 \\ b_2 \\ \vdots \\ b_n \end{pmatrix}$. Then $\overrightarrow{a} = \overrightarrow{b}$ if $m = n$ and $a_i = b_i$ for each $i \in \mathcal{I}_n$.

Example 1.3.3. Let $\overrightarrow{a} = (1, 3, y)$ and $\overrightarrow{b} = \begin{pmatrix} x^2 \\ 3 \\ 1 \end{pmatrix}$. Find $x, y \in \mathbb{R}$ such that $\overrightarrow{a} = \overrightarrow{b}$.

Solution. $\overrightarrow{a} = \overrightarrow{b}$ if and only if $x^2 = 1$ and $y = 1$. Hence, when $x = 1$ and $y = 1$ or $x = -1$ and $y = 1$, $\overrightarrow{a} = \overrightarrow{b}$. □

Exercises

1. Let $\overrightarrow{a} = \begin{pmatrix} x - 2y \\ 2x - y \\ 2z \end{pmatrix}$ and $\overrightarrow{b} = \begin{pmatrix} 2 \\ -2 \\ 1 \end{pmatrix}$. Find all $x, y, z \in \mathbb{R}$ such that $\overrightarrow{a} = \overrightarrow{b}$.
2. Let $\overrightarrow{a} = \begin{pmatrix} |x| \\ y^2 \end{pmatrix}$ and $\overrightarrow{b} = \begin{pmatrix} 1 \\ 4 \end{pmatrix}$. Find all $x, y \in \mathbb{R}$ such that $\overrightarrow{a} = \overrightarrow{b}$.
3. $\overrightarrow{a} = \begin{pmatrix} x - y \\ 4 \end{pmatrix}$ and $\overrightarrow{b} = \begin{pmatrix} 2 \\ x + y \end{pmatrix}$. Find all $x, y \in \mathbb{R}$ such that $\overrightarrow{a} - \overrightarrow{b}$ is a nonzero vector.

1.4 Operations of vectors

Definition 1.4.1. Let $\overrightarrow{a} = \begin{pmatrix} a_1 \\ a_2 \\ \vdots \\ a_n \end{pmatrix} \in \mathbb{R}^n$, $\overrightarrow{b} = \begin{pmatrix} b_1 \\ b_2 \\ \vdots \\ b_n \end{pmatrix} \in \mathbb{R}^n$ and $k \in \mathbb{R}$.

Then we define

Addition: $\overrightarrow{a} + \overrightarrow{b} = \begin{pmatrix} a_1 + b_1 \\ a_2 + b_2 \\ \vdots \\ a_n + b_n \end{pmatrix}$.

Substraction: $\overrightarrow{a} - \overrightarrow{b} = \begin{pmatrix} a_1 - b_1 \\ a_2 - b_2 \\ \vdots \\ a_n - b_n \end{pmatrix}$.

Scalar multiplication: $k\overrightarrow{a} = \begin{pmatrix} ka_1 \\ ka_2 \\ \vdots \\ ka_n \end{pmatrix}$.

Remark 1.4.2. The addition and subtraction of vectors are introducted in $\mathbb{R}^n$-space. If two vectors are in different spaces, say, one is in $\mathbb{R}^m$ and another in $\mathbb{R}^n$, where $m \neq n$, then the sum (or subtraction) of the two vectors are not defined. For example, $\begin{pmatrix} 1 \\ 2 \end{pmatrix} + \begin{pmatrix} 3 \\ 3 \\ 5 \end{pmatrix}$ is not defined.

Example 1.4.3. Let $\overrightarrow{a} = \begin{pmatrix} 1 \\ 2 \\ 0 \end{pmatrix}$ and $\overrightarrow{b} = \begin{pmatrix} 3 \\ 6 \\ 5 \end{pmatrix}$. Compute

(1) $\overrightarrow{a} + \overrightarrow{b}$; (2) $\overrightarrow{a} - \overrightarrow{b}$; (3) $5\overrightarrow{a}$; (4) $-\overrightarrow{a}$; (5) $2\overrightarrow{a} - 4\overrightarrow{b}$.

Solution. By Definition 1.4.1, we obtain

(1) $\overrightarrow{a} + \overrightarrow{b} = \begin{pmatrix} 1 \\ 2 \\ 0 \end{pmatrix} + \begin{pmatrix} 3 \\ 6 \\ 5 \end{pmatrix} = \begin{pmatrix} 1+3 \\ 2+6 \\ 0+5 \end{pmatrix} = \begin{pmatrix} 4 \\ 8 \\ 5 \end{pmatrix}$.

(2) $\overrightarrow{a} - \overrightarrow{b} = \begin{pmatrix} 1 \\ 2 \\ 0 \end{pmatrix} - \begin{pmatrix} 3 \\ 6 \\ 5 \end{pmatrix} = \begin{pmatrix} 1-3 \\ 2-6 \\ 0-5 \end{pmatrix} = \begin{pmatrix} -2 \\ -4 \\ -5 \end{pmatrix}$.

(3) $5\overrightarrow{a} = 5\begin{pmatrix} 1 \\ 2 \\ 0 \end{pmatrix} = \begin{pmatrix} (5)(1) \\ (5)(2) \\ (5)(0) \end{pmatrix} = \begin{pmatrix} 5 \\ 10 \\ 0 \end{pmatrix}$.

(4) $-\overrightarrow{a} = (-1)\begin{pmatrix} 1 \\ 2 \\ 0 \end{pmatrix} = \begin{pmatrix} (-1)(1) \\ (-1)(2) \\ (-1)(0) \end{pmatrix} = \begin{pmatrix} -1 \\ -2 \\ 0 \end{pmatrix}$.

(5) $2\overrightarrow{a} - 4\overrightarrow{b} = 2\begin{pmatrix} 1 \\ 2 \\ 0 \end{pmatrix} - 4\begin{pmatrix} 3 \\ 6 \\ 5 \end{pmatrix} = \begin{pmatrix} 2 \\ 4 \\ 0 \end{pmatrix} - \begin{pmatrix} 12 \\ 24 \\ 20 \end{pmatrix} = \begin{pmatrix} -10 \\ -20 \\ -20 \end{pmatrix}$.

□

By Definition 1.4.1, the following results on the operations of vectors can be easily proved, so we leave the proofs to reader.

Theorem 1.4.4. *Let* $\overrightarrow{a}, \overrightarrow{b}, \overrightarrow{c} \in \mathbb{R}^n$ *and let* $\alpha, \beta \in \mathbb{R}$. *Then*

(1) $\overrightarrow{a} + \overrightarrow{0} = \overrightarrow{a}$.

(2) $0\overrightarrow{a} = \overrightarrow{0}$.

(3) $\overrightarrow{a} + \overrightarrow{b} = \overrightarrow{b} + \overrightarrow{a}$.

(4) $(\overrightarrow{a} + \overrightarrow{b}) + \overrightarrow{c} = \overrightarrow{a} + (\overrightarrow{b} + \overrightarrow{c})$.

(5) $\alpha(\overrightarrow{a} + \overrightarrow{b}) = \alpha\overrightarrow{a} + \alpha\overrightarrow{b}$.

(6) $(\alpha + \beta)\overrightarrow{a} = \alpha\overrightarrow{a} + \beta\overrightarrow{a}$.

(7) $(\alpha\beta)\overrightarrow{a} = \alpha(\beta\overrightarrow{a})$.

The following examples give applications of Theorem 1.4.4.

Example 1.4.5. Let $\overrightarrow{a} = \begin{pmatrix} 1 \\ 2 \\ 0 \end{pmatrix}$, $\overrightarrow{b} = \begin{pmatrix} 3 \\ 6 \\ 5 \end{pmatrix}$ and $\overrightarrow{c} = \begin{pmatrix} 4 \\ 6 \\ 7 \end{pmatrix}$. Compute $\overrightarrow{a} - \overrightarrow{b} + 2\overrightarrow{c}$.

Solution. By Theorem 1.4.4, we have

$$\begin{aligned} \overrightarrow{a} - \overrightarrow{b} + 2\overrightarrow{c} &= (\overrightarrow{a} - \overrightarrow{b}) + 2\overrightarrow{c} = \left[\begin{pmatrix} 1 \\ 2 \\ 0 \end{pmatrix} - \begin{pmatrix} 3 \\ 6 \\ 5 \end{pmatrix}\right] + 2\begin{pmatrix} 4 \\ 6 \\ 7 \end{pmatrix} \\ &= \begin{pmatrix} -2 \\ -4 \\ -5 \end{pmatrix} + \begin{pmatrix} 8 \\ 12 \\ 14 \end{pmatrix} = \begin{pmatrix} 6 \\ 8 \\ 9 \end{pmatrix}. \end{aligned}$$

□

Example 1.4.6. Find $\overrightarrow{a}$ if $\frac{1}{5}[4\overrightarrow{a} - \begin{pmatrix} 1 \\ -2 \end{pmatrix}] = \begin{pmatrix} 0 \\ 5 \end{pmatrix} - 2\overrightarrow{a}$.

Solution. From the given equation, we see that $\overrightarrow{a} \in \mathbb{R}^2$ and

$$4\overrightarrow{a} - \begin{pmatrix} 1 \\ -2 \end{pmatrix} = 5[\begin{pmatrix} 0 \\ 5 \end{pmatrix} - 2\overrightarrow{a}] = \begin{pmatrix} 0 \\ 25 \end{pmatrix} - 10\overrightarrow{a}.$$

Hence, $4\vec{a} + 10\vec{a} = \begin{pmatrix} 0 \\ 25 \end{pmatrix} + \begin{pmatrix} 1 \\ -2 \end{pmatrix} = \begin{pmatrix} 1 \\ 23 \end{pmatrix}$. This implies $14\vec{a} = \begin{pmatrix} 1 \\ 23 \end{pmatrix}$ and $\vec{a} = \frac{1}{14}\begin{pmatrix} 1 \\ 23 \end{pmatrix} = \begin{pmatrix} \frac{1}{14} \\ \frac{23}{14} \end{pmatrix}$. □

Exercises

1. Let $\vec{a} = \begin{pmatrix} -1 \\ 2 \\ 3 \end{pmatrix}$, $\vec{b} = \begin{pmatrix} 2 \\ -2 \\ 0 \end{pmatrix}$, $\vec{c} = \begin{pmatrix} 3 \\ 0 \\ -1 \end{pmatrix}$ and $\alpha \in \mathbb{R}$.

   Compute

   a. $\vec{a} + \vec{b}$ b. $0\vec{a}$ c. $-2\vec{b}$ d. $2\vec{a} - 5\vec{b}$

   e. $2\vec{a} - \vec{b} + 5\vec{c}$ f. $4\vec{a} + \alpha\vec{b} - 2\vec{c}$.

2. Find x, y and z such that $\begin{pmatrix} 9 \\ 4y \\ 2z \end{pmatrix} + \begin{pmatrix} 3x \\ 8 \\ -6 \end{pmatrix} = \begin{pmatrix} 0 \\ 0 \\ 0 \end{pmatrix}$.

3. A company having 553 employees lists each employee's salary as a component of a vector $\vec{a}$ in $\mathbb{R}^{553}$. If a 6% salary increase has been approved, find the vector involving $\vec{a}$ that gives all the new salaries.

4. Let $\vec{a} = \begin{pmatrix} 110 \\ 88 \\ 40 \end{pmatrix}$ denote the current prices of three items at a store. Suppose that the store announces a sale so that the price of each item is reduced by 20%.

   (*a*) Find a 3-vector that gives the price changes for the three items.

   (*b*) Find a 3-vector that gives the new prices of the three items.

1.5 Linear combination

Definition 1.5.1. Let $m, n \in \mathbb{N}$. Let $\vec{a_1}, \vec{a_2}, \cdots, \vec{a_n}, \vec{b} \in \mathbb{R}^m$ and let $x_1, x_2, \cdots, x_n$ be n real numbers. If

$$\vec{b} = x_1\vec{a_1} + x_2\vec{a_2} + \cdots + x_n\vec{a_n}, \tag{1.1}$$

then $\vec{b}$ is called a linear combination of $\vec{a_1}, \vec{a_2}, \cdots, \vec{a_n}$.

Let

$$\overrightarrow{a_1} = \begin{pmatrix} a_{11} \\ a_{21} \\ \vdots \\ a_{m1} \end{pmatrix}, \overrightarrow{a_2} = \begin{pmatrix} a_{12} \\ a_{22} \\ \vdots \\ a_{m2} \end{pmatrix}, \cdots, \overrightarrow{a_m} = \begin{pmatrix} a_{1n} \\ a_{2n} \\ \vdots \\ a_{mn} \end{pmatrix}, \overrightarrow{b} = \begin{pmatrix} b_1 \\ b_2 \\ \vdots \\ b_m \end{pmatrix}.$$

Let

$$\overrightarrow{c} = x_1\overrightarrow{a_1} + x_2\overrightarrow{a_2} + \cdots + x_n\overrightarrow{a_n}.$$

Then by computation, we obtain

$$\begin{aligned} \overrightarrow{c} &= x_1\overrightarrow{a_1} + x_2\overrightarrow{a_2} + \cdots + x_n\overrightarrow{a_n} \\ &= x_1\begin{pmatrix} a_{11} \\ a_{21} \\ \vdots \\ a_{m1} \end{pmatrix} + x_2\begin{pmatrix} a_{12} \\ a_{22} \\ \vdots \\ a_{m2} \end{pmatrix} + \cdots + x_n\begin{pmatrix} a_{1n} \\ a_{2n} \\ \vdots \\ a_{mn} \end{pmatrix} \\ &= \begin{pmatrix} a_{11}x_1 \\ a_{21}x_1 \\ \vdots \\ a_{m1}x_1 \end{pmatrix} + \begin{pmatrix} a_{12}x_2 \\ a_{22}x_2 \\ \vdots \\ a_{m2}x_2 \end{pmatrix} + \cdots + \begin{pmatrix} a_{1m}x_n \\ a_{2m}x_n \\ \vdots \\ a_{mn}x_n \end{pmatrix} \\ &= \begin{pmatrix} a_{11}x_1 + a_{12}x_2 + \cdots + a_{1n}x_n \\ a_{21}x_1 + a_{22}x_2 + \cdots + a_{2n}x_n \\ \vdots \\ a_{m1}x_1 + a_{m2}x_2 + \cdots + a_{mn}x_n \end{pmatrix}. \end{aligned}$$

By Definition 1.3.1, $\overrightarrow{c} = \overrightarrow{b}$, that is,

$$\begin{cases} a_{11}x_1 + a_{12}x_2 + \cdots + a_{1n}x_n = b_1 \\ a_{21}x_1 + a_{22}x_2 + \cdots + a_{2n}x_n = b_2 \\ \vdots \\ a_{m1}x_1 + a_{m2}x_2 + \cdots + a_{mn}x_n = b_m. \end{cases} \tag{1.2}$$

Example 1.5.2. Let

$$\overrightarrow{a_1} = \begin{pmatrix} 1 \\ 0 \\ 2 \\ 4 \end{pmatrix}, \overrightarrow{a_2} = \begin{pmatrix} 2 \\ 3 \\ 1 \\ 2 \end{pmatrix}, \overrightarrow{a_3} = \begin{pmatrix} 0 \\ 1 \\ 3 \\ 1 \end{pmatrix}, \overrightarrow{a_4} = \begin{pmatrix} 0 \\ 0 \\ 1 \\ 0 \end{pmatrix} \text{ and } \overrightarrow{a_5} = \begin{pmatrix} 1 \\ 0 \\ 0 \\ 0 \end{pmatrix}.$$

Then

(*i*) Let $\vec{b} = \begin{pmatrix} -4 \\ -9 \\ 1 \\ 2 \end{pmatrix}$. Verify that $2\vec{a_1} - 3\vec{a_2} = \vec{b}$.

(*ii*) Let $\vec{b} = \begin{pmatrix} 5 \\ 7 \\ 7 \\ 9 \end{pmatrix}$. Verify that $\vec{a_1} + 2\vec{a_2} + \vec{a_3} = \vec{b}$.

(*iii*) Let $\vec{b} = \begin{pmatrix} -1 \\ -7 \\ 6 \\ 7 \end{pmatrix}$. Verify that $3\vec{a_1} - 2\vec{a_2} - \vec{a_3} + 5\vec{a_4} = \vec{b}$.

(*iv*) Let $\vec{b} = \begin{pmatrix} -5 \\ 1 \\ 2 \\ 4 \end{pmatrix}$. Verify that $\vec{a_1} + \vec{a_2} - 2\vec{a_3} + 5\vec{a_4} - 8\vec{a_5} = \vec{b}$.

Solution. (*i*) $2\vec{a_1} - 3\vec{a_2} = 2\begin{pmatrix} 1 \\ 0 \\ 2 \\ 4 \end{pmatrix} - 3\begin{pmatrix} 2 \\ 3 \\ 1 \\ 2 \end{pmatrix} = \begin{pmatrix} 2-6 \\ 0-9 \\ 4-3 \\ 8-6 \end{pmatrix} = \begin{pmatrix} -4 \\ -9 \\ 1 \\ 2 \end{pmatrix}$.

(*ii*) $\vec{a_1} + 2\vec{a_2} + \vec{a_3} = \begin{pmatrix} 1 \\ 0 \\ 2 \\ 4 \end{pmatrix} + 2\begin{pmatrix} 2 \\ 3 \\ 1 \\ 2 \end{pmatrix} + \begin{pmatrix} 0 \\ 1 \\ 3 \\ 1 \end{pmatrix} = \begin{pmatrix} 1+4+0 \\ 0+6+1 \\ 2+2+3 \\ 4+4+1 \end{pmatrix} = \begin{pmatrix} 5 \\ 7 \\ 7 \\ 9 \end{pmatrix}$.

(*iii*) Let $\vec{c} = 3\vec{a_1} - 2\vec{a_2} - \vec{a_3} + 5\vec{a_4}$. Then

$$\vec{c} = 3\begin{pmatrix} 1 \\ 0 \\ 2 \\ 4 \end{pmatrix} - 2\begin{pmatrix} 2 \\ 3 \\ 1 \\ 2 \end{pmatrix} - \begin{pmatrix} 0 \\ 1 \\ 3 \\ 1 \end{pmatrix} + 5\begin{pmatrix} 0 \\ 0 \\ 1 \\ 0 \end{pmatrix} = \begin{pmatrix} -1 \\ -7 \\ 6 \\ 7 \end{pmatrix}.$$

(*iv*) Let $\vec{c} = \vec{a_1} + \vec{a_2} - 2\vec{a_3} + 5\vec{a_4} - 8\vec{a_5}$. Then

$$\vec{c} = \begin{pmatrix} 1 \\ 0 \\ 2 \\ 4 \end{pmatrix} + \begin{pmatrix} 2 \\ 3 \\ 1 \\ 2 \end{pmatrix} - 2\begin{pmatrix} 0 \\ 1 \\ 3 \\ 1 \end{pmatrix} + 5\begin{pmatrix} 0 \\ 0 \\ 1 \\ 0 \end{pmatrix} - 8\begin{pmatrix} 1 \\ 0 \\ 0 \\ 0 \end{pmatrix} = \begin{pmatrix} -5 \\ 1 \\ 2 \\ 4 \end{pmatrix}.$$

□

Example 1.5.3. Let $\vec{a_1} = \begin{pmatrix} 1 \\ 2 \end{pmatrix}$, $\vec{a_2} = \begin{pmatrix} 2 \\ 1 \end{pmatrix}$. Show that $\vec{b} = \begin{pmatrix} 8 \\ 7 \end{pmatrix}$ is a linear combination of $\vec{a_1}$ and $\vec{a_2}$.

Solution. Let $x_1\overrightarrow{a_1} + x_2\overrightarrow{a_2} = \overrightarrow{b}$. Then

$$x_1\begin{pmatrix}1\\2\end{pmatrix} + x_2\begin{pmatrix}2\\1\end{pmatrix} = \begin{pmatrix}8\\7\end{pmatrix}.$$

This implies $\begin{pmatrix}x_1 + 2x_2\\2x_1 + x_2\end{pmatrix} = \begin{pmatrix}8\\7\end{pmatrix}$. By Definition 1.3.1, we have

$$\begin{cases}x_1 + 2x_2 = 8\\2x_1 + x_2 = 7.\end{cases}$$

Solving the system, we get $x_1 = 2$ and $x_2 = 3$. Hence, $\overrightarrow{b} = 2\overrightarrow{a_1} + 3\overrightarrow{a_2}$, so $\overrightarrow{b}$ is a linear combination of $\overrightarrow{a_1}$ and $\overrightarrow{a_2}$. □

Similarly, the equality of two row vectors and the three operations: addition, substraction and scalar multiplication for row vectors can be defined in a similar fashion.

Example 1.5.4. Let $\overrightarrow{a} = (0, 1, 2, 4)$ and $\overrightarrow{b} = (3, 2, 1, 0)$. Compute
(*i*) $\overrightarrow{a} + \overrightarrow{b}$; (*ii*) $\overrightarrow{a} - \overrightarrow{b}$; (*iii*) $2\overrightarrow{a} + 3\overrightarrow{b}$.

Solution. (*i*) $\overrightarrow{a} + \overrightarrow{b} = (0, 1, 2, 4) + (3, 2, 1, 0) = (3, 3, 3, 4)$.
(*ii*) $\overrightarrow{a} - \overrightarrow{b} = (0, 1, 2, 4) - (3, 2, 1, 0) = (-3, -1, 1, 4)$.
(*iii*) $2\overrightarrow{a} + 3\overrightarrow{b} = (0, 2, 4, 8) + (9, 6, 3, 0) = (9, 8, 7, 8)$. □

Exercises

1. Let $\overrightarrow{a_1} = \begin{pmatrix}1\\2\end{pmatrix}$, $\overrightarrow{a_2} = \begin{pmatrix}1\\1\end{pmatrix}$ and $\overrightarrow{b} = \begin{pmatrix}1\\0\end{pmatrix}$. Show whether $\overrightarrow{b}$ is a linear combination of $\overrightarrow{a_1}$ and $\overrightarrow{a_2}$.

2. Let $\overrightarrow{a_1} = \begin{pmatrix}1\\0\\1\end{pmatrix}$, $\overrightarrow{a_2} = \begin{pmatrix}1\\3\\2\end{pmatrix}$ and $\overrightarrow{a_3} = \begin{pmatrix}-1\\1\\3\end{pmatrix}$.

   (*i*) Let $\overrightarrow{b} = \begin{pmatrix}-1\\-9\\-4\end{pmatrix}$. Verify whether $2\overrightarrow{a_1} - 3\overrightarrow{a_2} = \overrightarrow{b}$.

   (*ii*) Let $\overrightarrow{b} = \begin{pmatrix}3\\6\\6\end{pmatrix}$. Verify whether $\overrightarrow{a_1} - 2\overrightarrow{a_2} = \overrightarrow{b}$.

   (*iii*) Let $\overrightarrow{b} = \begin{pmatrix}2\\7\\7\end{pmatrix}$. Verify whether $\overrightarrow{a_1} + 2\overrightarrow{a_2} + \overrightarrow{a_3} = \overrightarrow{b}$.

3. Let $\overrightarrow{a_1} = \begin{pmatrix} 1 \\ 1 \end{pmatrix}$ and $\overrightarrow{a_2} = \begin{pmatrix} 2 \\ 1 \end{pmatrix}$. Show whether $\overrightarrow{b} = \begin{pmatrix} 0 \\ 1 \end{pmatrix}$ is a linear combination of $\overrightarrow{a_1}$ and $\overrightarrow{a_2}$.

1.6 The dot product of two vectors in $\mathbb{R}^n$

Definition 1.6.1. Let $\overrightarrow{a} = \begin{pmatrix} a_1 \\ a_2 \\ \vdots \\ a_n \end{pmatrix} \in \mathbb{R}^n$ and $\overrightarrow{b} = \begin{pmatrix} b_1 \\ b_2 \\ \vdots \\ b_n \end{pmatrix} \in \mathbb{R}^n$. The dot product of $\overrightarrow{a}$ and $\overrightarrow{b}$, denoted by $\overrightarrow{a} \cdot \overrightarrow{b}$, is defined by

$$\overrightarrow{a} \cdot \overrightarrow{b} = a_1b_1 + a_2b_2 + \cdots + a_nb_n. \tag{1.3}$$

It is convenient to write the dot product of $\overrightarrow{a}$ and $\overrightarrow{b}$ in the following form:

$$\overrightarrow{a} \cdot \overrightarrow{b} = (a_1, a_2, \cdots, a_n) \begin{pmatrix} b_1 \\ b_2 \\ \vdots \\ n_n \end{pmatrix} = a_1b_1 + a_2b_2 + \cdots + a_nb_n. \tag{1.4}$$

In words, $\overrightarrow{a} \cdot \overrightarrow{b}$ is the sum of the products of the corresponding components of $\overrightarrow{a}$ and $\overrightarrow{b}$.

Example 1.6.2. Let $\overrightarrow{a} = \begin{pmatrix} 1 \\ 0 \\ 4 \\ 2 \\ 0 \end{pmatrix}$ and $\overrightarrow{b} = \begin{pmatrix} 2 \\ 5 \\ 0 \\ 1 \\ 4 \end{pmatrix}$. Calculate $\overrightarrow{a} \cdot \overrightarrow{b}$.

Solution.

$$\overrightarrow{a} \cdot \overrightarrow{b} = (1, 0, 4, 2, 0) \begin{pmatrix} 2 \\ 5 \\ 0 \\ 1 \\ 4 \end{pmatrix} = (1)(2) + (0)(5) + (4)(0) + (2)(1) + (0)(4) = 4.$$

□

Example 1.6.3. Let $\overrightarrow{u} = (-1, 3, 5, -7)$ and $\overrightarrow{v} = (5, -4, 7, 0)$. Find $\overrightarrow{u} \cdot \overrightarrow{v}$.

Solution. $\overrightarrow{u} \cdot \overrightarrow{v} = (-1) \times 5 + 3 \times (-4) + 5 \times 7 + (-7) \times 0 = 18.$ □

Example 1.6.4. Let $\overrightarrow{a} = \begin{pmatrix} x^2 \\ 1 \end{pmatrix}$ and $\overrightarrow{b} = \begin{pmatrix} 2 \\ -8 \end{pmatrix}$. Find $x \in \mathbb{R}$ such that $\overrightarrow{a} \cdot \overrightarrow{b} = 0$.

Solution. $\overrightarrow{a} \cdot \overrightarrow{b} = (x^2, 1)\begin{pmatrix} 2 \\ -8 \end{pmatrix} = x^2(2) + (1)(-8) = 2x^2 - 8 = 0$. This implies $x^2 = 4$. Solving the equation, we have $x = 2$ or $x = -2$. Hence, when $x = 2$ or $x = -2$, the dot product $\overrightarrow{a} \cdot \overrightarrow{b} = 0$. □

The following result follows directly from the definition of the dot product.

Theorem 1.6.5. *Let $\overrightarrow{a}, \overrightarrow{b}, \overrightarrow{c} \in \mathbb{R}^n$ and let $\alpha, \beta \in \mathbb{R}$. Then*

(1) $\overrightarrow{a} \cdot \overrightarrow{0} = 0$.

(2) $\overrightarrow{a} \cdot \overrightarrow{b} = \overrightarrow{b} \cdot \overrightarrow{a}$. *(commutative law for scalar product)*

(3) $\overrightarrow{a} \cdot (\overrightarrow{b} + \overrightarrow{c}) = \overrightarrow{a} \cdot \overrightarrow{b} + \overrightarrow{a} \cdot \overrightarrow{c}$. *(distributive law for scalar product)*

(4) $(\alpha \overrightarrow{a}) \cdot \overrightarrow{b} = \overrightarrow{a} \cdot (\alpha \overrightarrow{b})$.

Example 1.6.6. Let $\overrightarrow{a} = \begin{pmatrix} 1 \\ 0 \\ 4 \\ 2 \end{pmatrix}$, $\overrightarrow{b} = \begin{pmatrix} 2 \\ 5 \\ 0 \\ 1 \end{pmatrix}$ and $\overrightarrow{c} = \begin{pmatrix} 1 \\ 2 \\ 0 \\ 1 \end{pmatrix}$. Calculate $\overrightarrow{a} \cdot (\overrightarrow{b} + \overrightarrow{c})$.

Solution.

$$\begin{aligned} \overrightarrow{a} \cdot (\overrightarrow{b} + \overrightarrow{c}) &= \overrightarrow{a} \cdot \overrightarrow{b} + \overrightarrow{a} \cdot \overrightarrow{c} = (1,0,4,2)\begin{pmatrix} 2 \\ 5 \\ 0 \\ 1 \end{pmatrix} + (1,0,4,2)\begin{pmatrix} 1 \\ 2 \\ 0 \\ 1 \end{pmatrix} \\ &= 4 + 3 = 7. \end{aligned}$$

□

Exercises

1. Let $\overrightarrow{a} = \begin{pmatrix} -1 \\ 2 \\ 3 \end{pmatrix}$, $\overrightarrow{b} = \begin{pmatrix} 2 \\ -2 \\ 0 \end{pmatrix}$, $\overrightarrow{c} = \begin{pmatrix} 3 \\ 0 \\ -1 \end{pmatrix}$ and $\alpha \in \mathbb{R}$.

   Compute

   (a). $\overrightarrow{a} \cdot \overrightarrow{b}$; (b). $\overrightarrow{a} \cdot \overrightarrow{c}$; (c). $\overrightarrow{b} \cdot \overrightarrow{c}$; (d). $\overrightarrow{a} \cdot (\overrightarrow{b} + \overrightarrow{c})$.

2. Let $\overrightarrow{u} = (1, -1, 0, 2)$, $\overrightarrow{v} = (-2, -1, 1, -4)$ and $\overrightarrow{w} = (3, 2, -1, 0)$.

   (a) Find a vector $\overrightarrow{a} \in \mathbb{R}^4$ such that $2\overrightarrow{u} - 3\overrightarrow{v} - \overrightarrow{a} = \overrightarrow{w}$.

   (b) Compute $\overrightarrow{u} \cdot \overrightarrow{v}$, $\overrightarrow{u} \cdot \overrightarrow{w}$ and $\overrightarrow{w} \cdot \overrightarrow{v}$.

   (c) Find a vector $\overrightarrow{a}$ such that

$$\frac{1}{2}[2\overrightarrow{u} - 3\overrightarrow{v} + \overrightarrow{a}] = 2\overrightarrow{w} + \overrightarrow{u} - 2\overrightarrow{a}.$$

3. Assume that the percentages for homework, test 1, test 2 and final exam for a course are 10%, 25%, 25%, 40%, respectively. The total marks for homework, test 1, test 2 and final exam are 10, 50, 50, 90, respectively. A student's corresponding marks are 8, 46, 48, 81, respectively. What are the student's final marks out of 100.

Chapter 2

Matrices

2.1 Definition of a matrix

Definition 2.1.1. An $m \times n$ matrix A is a rectangular array of mn numbers arranged in m rows and n columns:

$$A = \begin{pmatrix} a_{11} & a_{12} & \cdots & a_{1n} \\ a_{21} & a_{22} & \cdots & a_{2n} \\ \vdots & \vdots & \ddots & \vdots \\ a_{m1} & a_{m2} & \cdots & a_{mn} \end{pmatrix}. \tag{2.1}$$

The ijth component of A, denoted a_{ij}, is the number appearing in the ith row and jth column of A. Sometimes, it is useful to write $A = (a_{ij})$. Capital letters are usually applied to denote matrices. $m \times n$ is called the size of A. a_{ij} is called an (i, j)-entry of A, or entry or element of A if there are no confusions.

Example 2.1.2. 1. Let $A = (2)$. Then the size of A is 1×1.

2. Let $A = \begin{pmatrix} 1 & 2 \\ 3 & 4 \end{pmatrix}$. Then the size of A is 2×2.

3. Let $B = \begin{pmatrix} 1 & 2 & 0 \\ 3 & 4 & -1 \end{pmatrix}$. Then the size of B is 2×3.

4. Let $C = \begin{pmatrix} 0 & 0 & 0 \\ 0 & 0 & 0 \\ 0 & 0 & 0 \\ 0 & 0 & 0 \end{pmatrix}$. Then the size of C is 4×3.

Zero matrices An $m \times n$ matrix is called a zero matrix if all entries of A are zero.

Sometimes, we denote by 0 a zero matrix if there is no confusion.

Example 2.1.3. The following are zero matrices.

$$A = (0),\ B = \begin{pmatrix} 0 & 0 & 0 \\ 0 & 0 & 0 \end{pmatrix} \text{ and } C = \begin{pmatrix} 0 & 0 & 0 \\ 0 & 0 & 0 \\ 0 & 0 & 0 \\ 0 & 0 & 0 \end{pmatrix}.$$

Row matrices A $1 \times n$ matrix is called a row matrix.

Example 2.1.4. (1) $A = \begin{pmatrix} 0 & 0 & 0 \end{pmatrix}$ is a 1×3 row matrix.
(2) $A = \begin{pmatrix} 1 & 3 & 5 & 7 \end{pmatrix}$ is a 1×4 row matrix.

A $1 \times n$ matrix $\begin{pmatrix} a_1 & a_2 & \cdots & a_n \end{pmatrix}$ can be treated as a row vector.

Column matrices An $n \times 1$ matrix is called a column matrix.

Example 2.1.5. $A = \begin{pmatrix} 0 \\ 0 \\ 0 \\ 0 \end{pmatrix}$ is a 4×1 column matrix and $A = \begin{pmatrix} 2 \\ 1 \\ 3 \end{pmatrix}$ is a 3×1 column matrix.

We can treat an $n \times 1$ matrix as a column vector. Let

$$\overrightarrow{a_1} = \begin{pmatrix} a_{11} \\ a_{21} \\ \vdots \\ a_{m1} \end{pmatrix}, \quad \overrightarrow{a_2} = \begin{pmatrix} a_{12} \\ a_{22} \\ \vdots \\ a_{m2} \end{pmatrix}, \cdots, \overrightarrow{a_n} = \begin{pmatrix} a_{1n} \\ a_{2n} \\ \vdots \\ a_{mn} \end{pmatrix}. \tag{2.2}$$

The vectors $\overrightarrow{a_1}$, $\overrightarrow{a_2}$, $\cdots$, $\overrightarrow{a_n}$ are called column vectors of the matrix A. We can rewrite the matrix A as

$$A = (\overrightarrow{a_1}\,\overrightarrow{a_2}\,\cdots\,\overrightarrow{a_n}). \tag{2.3}$$

Let

$$\begin{aligned} \overrightarrow{r_1} &= (a_{11}, a_{12}, \cdots, a_{1n}), \\ \overrightarrow{r_2} &= (a_{21}, a_{22}, \cdots, a_{2n}), \\ &\vdots \\ \overrightarrow{r_m} &= (a_{m1}, a_{m2}, \cdots, a_{mn}). \end{aligned} \tag{2.4}$$

The vectors $\overrightarrow{r_1}$, $\overrightarrow{r_2}$, $\cdots$, $\overrightarrow{r_m}$ are called row vectors of A. We can rewrite the matrix A as

$$A = \begin{pmatrix} \overrightarrow{r_1} \\ \overrightarrow{r_2} \\ \vdots \\ \overrightarrow{r_m} \end{pmatrix}. \tag{2.5}$$

Note that $\overrightarrow{a_1}$, $\overrightarrow{a_2}$, $\cdots$, $\overrightarrow{a_n}$ are in $\mathbb{R}^m$ while $\overrightarrow{r_1}$, $\overrightarrow{r_2}$, $\cdots$, $\overrightarrow{r_m}$ are in $\mathbb{R}^n$.

Example 2.1.6. Let $A = \begin{pmatrix} 1 & 0 & 0 & 1 \\ 2 & 2 & 1 & 1 \\ 0 & 1 & 2 & 4 \end{pmatrix}$. Use column vectors and row vectors of A to rewrite A

Solution. Let $\overrightarrow{a_1} = \begin{pmatrix} 1 \\ 2 \\ 0 \end{pmatrix}$, $\overrightarrow{a_2} = \begin{pmatrix} 0 \\ 2 \\ 1 \end{pmatrix}$, $\overrightarrow{a_3} = \begin{pmatrix} 0 \\ 1 \\ 2 \end{pmatrix}$ and $\overrightarrow{a_4} = \begin{pmatrix} 1 \\ 1 \\ 4 \end{pmatrix}$.
Then A can be rewritten as $A = (\overrightarrow{a_1}\,\overrightarrow{a_2}\,\overrightarrow{a_3}\,\overrightarrow{a_4})$.

Let $\overrightarrow{r_1} = (1,0,0,1)$, $\overrightarrow{r_2} = (2,2,1,1)$ and $\overrightarrow{r_1} = (0,1,2,4)$. Then A can be rewritten as $A = \begin{pmatrix} \overrightarrow{r_1} \\ \overrightarrow{r_2} \\ \overrightarrow{r_3} \end{pmatrix}$. □

Exercises

1. Find the sizes of the following matrices:

   (i) $A = (2)$; (ii) $B = \begin{pmatrix} 5 & 8 \\ 1 & 4 \end{pmatrix}$; (iii) $C = \begin{pmatrix} 10 & 3 & 8 \\ 10 & 3 & 4 \end{pmatrix}$;

   (iv) $D = \begin{pmatrix} 1 & 0 & 0 \\ 0 & 10 & 2 \\ 0 & 0 & 1 \\ 6 & 9 & 10 \end{pmatrix}$.

2. Use column vectors and row vectors to rewrite each of the following matrices.

   $$A = \begin{pmatrix} 3 & 3 & 2 \\ 1 & 8 & 1 \\ 5 & 4 & 10 \end{pmatrix}; B = \begin{pmatrix} 1 & 9 & 7 & 2 \\ 3 & 10 & 8 & 7 \\ 4 & 3 & 10 & 4 \end{pmatrix}; C = \begin{pmatrix} 9 & 8 & 7 \\ 6 & 5 & 4 \\ 3 & 2 & 1 \\ 0 & -1 & -2 \end{pmatrix}.$$

2.2 Square matrices

An $n \times n$ matrix is called a square matrix of order n, that is,

$$A = \begin{pmatrix} a_{11} & a_{12} & \cdots & a_{1n} \\ a_{21} & a_{22} & \cdots & a_{2n} \\ \vdots & \vdots & \ddots & \vdots \\ a_{n1} & a_{n2} & \cdots & a_{nn} \end{pmatrix}. \tag{2.6}$$

The entries $a_{11}, a_{22}, \cdots, a_{nn}$ are said to be on the main diagonal of A. The sum of these entries, denote by $\text{tr}(A)$, is called the trace of A, that is,

$$\text{tr}(A) = \sum_{i=1}^{n} a_{ii} = a_{11} + a_{22} + \cdots + a_{nn}. \tag{2.7}$$

Example 2.2.1. Let $A = \begin{pmatrix} 15 & 21 \\ 44 & 25 \end{pmatrix}$. Then A is a square matrix of order 2, the numbers 15, 25 are on the main diagonal of A and $\mathrm{tr}(A) = 15 + 25 = 40$.

Symmetric matrices A square matrix is said to be symmetric if

$$a_{ij} = a_{ji} \quad \text{for } i, j \in \mathcal{I}_n.$$

We can write the symmetric matrix in an explicit form

$$A = \begin{pmatrix} a_{11} & a_{12} & \cdots & a_{1n} \\ a_{12} & a_{22} & \cdots & a_{2n} \\ \vdots & \vdots & \ddots & \vdots \\ a_{1n} & a_{2n} & \cdots & a_{nn} \end{pmatrix}.$$

From the above matrix, we see that A is symmetric if the ith row and the ith column are same for each $i \in \mathcal{I}_n$.

Example 2.2.2. (1) The following matrices are symmetric.

$$\begin{pmatrix} 7 & -3 \\ -3 & 0 \end{pmatrix} \quad \begin{pmatrix} 1 & 4 & 5 \\ 4 & -3 & 0 \\ 5 & 0 & 7 \end{pmatrix} \quad \begin{pmatrix} 1 & x^2 & 2 \\ x^2 & 0 & x \\ 2 & x & 3 \end{pmatrix}$$

(2) The following matrices are not symmetric.

$$\begin{pmatrix} 7 & -3 \\ 2 & 0 \end{pmatrix} \quad \begin{pmatrix} 1 & 4 & 5 \\ 2 & -3 & 0 \\ 5 & 0 & 7 \end{pmatrix} \quad \begin{pmatrix} 2 & x^2+1 & 2 \\ 0 & 1 & 2 \\ 2 & 2 & 2 \end{pmatrix}$$

Triangular matrices A square matrix is said to be
(*i*) **lower triangular** if all the entries above the main diagonal are zero;
(*ii*) **upper triangular** if all the entries below the main diagonal are zero;
(*iii*) **diagonal** if it is lower and upper triangular;
(*iv*) **triangular** if it is lower triangular or upper triangular.

Example 2.2.3. (1) The following matrices are lower triangular.

$$\begin{pmatrix} 0 & 0 \\ 2 & 0 \end{pmatrix} \quad \begin{pmatrix} 1 & 0 & 0 \\ 2 & 0 & 0 \\ 0 & 1 & 1 \end{pmatrix} \quad \begin{pmatrix} 0 & 0 & 0 \\ 0 & 0 & 0 \\ 0 & 0 & 0 \end{pmatrix} \quad \begin{pmatrix} 1 & 0 & 0 \\ 1 & 0 & 0 \\ x & 0 & 0 \end{pmatrix}$$

(2) The following matrices are upper triangular.

$$\begin{pmatrix} 0 & 1 \\ 0 & 0 \end{pmatrix} \quad \begin{pmatrix} 1 & 0 & 1 \\ 0 & 0 & 0 \\ 0 & 0 & 1 \end{pmatrix} \quad \begin{pmatrix} 0 & 0 & 0 \\ 0 & 0 & 0 \\ 0 & 0 & 0 \end{pmatrix}$$

(3) The following matrices are diagonal.

$$\begin{pmatrix} 1 & 0 \\ 0 & 2 \end{pmatrix} \quad \begin{pmatrix} 1 & 0 & 0 \\ 0 & 0 & 0 \\ 0 & 0 & 1 \end{pmatrix} \quad \begin{pmatrix} 0 & 0 & 0 \\ 0 & 0 & 0 \\ 0 & 0 & 0 \end{pmatrix}$$

(4) The following matrices are not triangular matrices.

$$\begin{pmatrix} 1 & 0 & 0 \\ 0 & 2 & 0 \\ 1 & 1 & 1 \end{pmatrix} \quad \begin{pmatrix} 1 & 0 & 1 \\ 2 & 0 & 0 \\ 0 & 0 & 1 \end{pmatrix} \quad \begin{pmatrix} 0 & 0 & 0 \\ 0 & 0 & 0 \end{pmatrix}$$

Identity matrix A diagonal matrix is said to be an identity matrix if all of its entries on the main diagonal are 1.

We denote by I or I_n an $n \times n$ identity matrix. Hence,

$$I_n = \begin{pmatrix} 1 & 0 & \cdots & 0 \\ 0 & 1 & \cdots & 0 \\ \vdots & \vdots & \ddots & \vdots \\ 0 & 0 & \cdots & 1 \end{pmatrix}$$

Example 2.2.4. The following are identity matrices.

$$I_1 = (1), \quad I_2 = \begin{pmatrix} 1 & 0 \\ 0 & 1 \end{pmatrix}, \quad I_3 = \begin{pmatrix} 1 & 0 & 0 \\ 0 & 1 & 0 \\ 0 & 0 & 1 \end{pmatrix}, \quad I_4 = \begin{pmatrix} 1 & 0 & 0 & 0 \\ 0 & 1 & 0 & 0 \\ 0 & 0 & 1 & 0 \\ 0 & 0 & 0 & 1 \end{pmatrix}.$$

Exercises

1. Let

$$A_1 = \begin{pmatrix} 0 & -1 \\ -1 & 0 \end{pmatrix}, A_2 = \begin{pmatrix} 1 & 4 & 5 \\ 4 & -3 & -1 \\ 5 & 0 & 7 \end{pmatrix}, A_3 = \begin{pmatrix} y & x^3 & 1 \\ x^3 & y & x \\ 1 & x & z \end{pmatrix}.$$

   (i) Find the trace of each of the above matrices.

   (ii) Determine which of the above matrices are symmetric.

2. Identify whether the given matrix is a lower triangular, upper triangular, diagonal, triangular or identity matrix.

$$A_1 = \begin{pmatrix} 0 & 2 \\ 0 & 0 \end{pmatrix} \quad A_2 = \begin{pmatrix} 1 & 0 & 0 \\ 2 & 0 & 0 \\ 0 & 0 & 1 \end{pmatrix} \quad A_3 = \begin{pmatrix} 0 & 0 & 0 \\ 0 & 0 & 0 \\ 0 & 0 & 0 \end{pmatrix}$$

$$A_4 = \begin{pmatrix} 1 & 0 & x \\ 1 & 0 & 0 \\ x & 0 & 0 \end{pmatrix} \quad A_5 = \begin{pmatrix} 0 & 1 \\ 1 & 0 \end{pmatrix} \quad A_6 = \begin{pmatrix} 1 & 0 & 1 \\ 0 & 0 & 0 \\ 0 & 0 & 1 \end{pmatrix}$$

$$A_7 = \begin{pmatrix} 2 & 0 \\ 0 & 2 \end{pmatrix} \quad A_8 = \begin{pmatrix} 1 & 0 & 0 \\ 0 & 0 & 0 \\ 2 & 0 & 1 \end{pmatrix} \quad A_9 = \begin{pmatrix} 1 & 0 & 0 \\ 0 & 2 & 0 \\ -1 & 1 & 1 \end{pmatrix}$$

$$A_{10} = \begin{pmatrix} 1 & 0 & 1 \\ 2 & 0 & 0 \\ 1 & 0 & 1 \end{pmatrix} \quad A_{11} = \begin{pmatrix} 0 & 0 \\ 0 & 0 \\ 0 & 0 \end{pmatrix} \quad A_{12} = \begin{pmatrix} 0 & 0 & 1 \\ 0 & 1 & 0 \\ 1 & 0 & 0 \end{pmatrix}$$

2.3 Row echelon matrices

Leading entries A row in an $m \times n$ matrix is said to be a zero row if all the entries in that row are zero. If one of entries in a row is nonzero, then it is called a nonzero row.

Example 2.3.1. In the matrix $\begin{pmatrix} 1 & 0 & 0 \\ 2 & 1 & 0 \\ 0 & 0 & 0 \end{pmatrix}$, the first two rows are nonzero rows and the last row is a zero row.

The first nonzero entry (starting from the left) in a nonzero row is called the **leading entry** of the row. If a leading entry is 1, it is called a **leading** 1.

Example 2.3.2. Let $A = \begin{pmatrix} 1 & 6 & 0 & 4 \\ 0 & 2 & 0 & 5 \\ 3 & 0 & 0 & 10 \\ 0 & 0 & 0 & 0 \end{pmatrix}$. Find the leading entries of A.

Solution. The numbers 1, 2 and 3 are leading entries of A. □

Row echelon matrices

An $m \times n$ matrix is said to be a row echelon matrix if it satisfies the following two conditions:

(1) Each nonzero row (if any) lies above every zero row, that is, all zero rows appear at the bottom of the matrix.

(2) For any two successive nonzero rows, the leading entry in the lower row occurs further to the right than the leading entry in the higher row.

Example 2.3.3. (1) The following matrices are row echelon matrices.

$$\begin{pmatrix} 1 & 0 \\ 0 & 1 \end{pmatrix}, \begin{pmatrix} 2 & 0 & 0 & 3 & 5 \\ 0 & 0 & 0 & 1 & 2 \end{pmatrix}, \begin{pmatrix} 2 & 0 & 1 \\ 0 & 1 & 0 \\ 0 & 0 & 1 \end{pmatrix}, \begin{pmatrix} 1 & 0 & 0 & 6 & 0 \\ 0 & 0 & 2 & 3 & 0 \\ 0 & 0 & 0 & 0 & 1 \\ 0 & 0 & 0 & 0 & 0 \end{pmatrix}.$$

(2) The following matrices are not row echelon matrices.

$$\begin{pmatrix} 1 & 0 & 0 \\ 0 & 0 & 1 \\ 0 & 1 & 0 \end{pmatrix}, \begin{pmatrix} 0 & 0 & 1 \\ 0 & 0 & 0 \\ 0 & 0 & 1 \end{pmatrix}, \begin{pmatrix} 0 & 0 & 1 & 0 \\ 1 & 0 & 0 & 0 \\ 0 & 0 & 0 & 0 \end{pmatrix}, \begin{pmatrix} 1 & 0 & 0 & 0 \\ 0 & 0 & 0 & 0 \\ 0 & 0 & 0 & 1 \end{pmatrix}.$$

It is clear that if a column contains a leading entry of some row, then all entries below the leading entry of the column must be zero.

Example 2.3.4. In the matrix $\begin{pmatrix} 1 & 0 & 1 & 0 \\ 0 & 5 & 0 & 0 \\ 0 & 3 & 0 & 1 \\ 0 & 0 & 0 & 0 \end{pmatrix}$, the column 2 contains the leading entry 5 of the second row and there is an entry 3 in the column 2 which is below 5. Hence the matrix is not a row echelon matrix.

Reduced row echelon matrices

An $m \times n$ row echelon matrix is said to be a reduced row echelon matrix if it satisfies the following two conditions:

(3) Each leading entry is 1.

(4) For each column containing a leading 1, all entries except the leading one are zero.

Example 2.3.5. (1) The following are reduced row echelon matrices:

$$\begin{pmatrix} 1 & 0 & 0 & 1 \\ 0 & 1 & 0 & 0 \\ 0 & 0 & 1 & 0 \\ 0 & 0 & 0 & 0 \end{pmatrix}, \begin{pmatrix} 1 & 0 & 0 & 6 & 0 \\ 0 & 0 & 1 & 3 & 0 \\ 0 & 0 & 0 & 0 & 1 \\ 0 & 0 & 0 & 0 & 0 \end{pmatrix}, \begin{pmatrix} 0 & 1 & 0 & 0 & 1 \\ 0 & 0 & 0 & 1 & 1 \end{pmatrix}.$$

(2) The following are row echelon matrices, but not reduced row echelon matrices:

$$\begin{pmatrix} 2 & 0 & 1 \\ 0 & 1 & 0 \\ 0 & 0 & 1 \end{pmatrix}, \begin{pmatrix} 1 & 0 & 2 & 6 & 0 \\ 0 & 0 & 1 & 3 & 0 \\ 0 & 0 & 0 & 0 & 1 \\ 0 & 0 & 0 & 0 & 0 \end{pmatrix}, \begin{pmatrix} 1 & 0 & 0 & 2 & 0 \\ 0 & 0 & 2 & 1 & 0 \end{pmatrix}.$$

Solution. We only consider (2). These matrices are row echelon matrices. However, the first matrix contains the leading entry 2 which is not 1, the second matrix contains the column 3 where there is a nonzero entry 2 except the leading 1 and the third matrix has a leading entry which is not 1. Hence, they are all not reduced row echelon matrices. □

Exercises

1. Which of the following matrices are row echelon matrices?

$$A = \begin{pmatrix} 1 & 0 \\ 0 & 1 \end{pmatrix} \quad B = \begin{pmatrix} 9 & 4 & 3 & 5 & 9 \\ 0 & 0 & 0 & 7 & 4 \end{pmatrix} \quad C = \begin{pmatrix} 2 & 5 & 0 \\ 0 & 9 & 0 \\ 0 & 0 & 8 \end{pmatrix}$$

$$D = \begin{pmatrix} 3 & 0 & 1 & 6 & 0 \\ 0 & 0 & 8 & 8 & 6 \\ 0 & 0 & 0 & 0 & 1 \\ 0 & 0 & 0 & 0 & 0 \end{pmatrix} \quad E = \begin{pmatrix} 1 & 0 & 0 \\ 0 & 0 & 1 \\ 0 & 1 & 0 \end{pmatrix} \quad F = \begin{pmatrix} 1 & 4 & 1 \\ 0 & 0 & 0 \\ 0 & 6 & 6 \end{pmatrix}$$

$$G = \begin{pmatrix} 0 & 8 & 1 & 7 \\ 2 & 9 & 8 & 0 \\ 0 & 0 & 0 & 0 \end{pmatrix} \quad H = \begin{pmatrix} 0 & 0 & 0 & 0 \\ 4 & 0 & 5 & 8 \\ 0 & 9 & 6 & 1 \end{pmatrix}$$

2. Which of the following matrices are reduced row echelon matrices?

$$A = \begin{pmatrix} 1 & 0 & 0 & 1 \\ 0 & 1 & 0 & 1 \\ 0 & 0 & 1 & 0 \\ 0 & 0 & 0 & 0 \end{pmatrix} \quad B = \begin{pmatrix} 1 & 5 & 0 & 6 & 0 \\ 0 & 0 & 1 & 3 & 0 \\ 0 & 0 & 0 & 0 & 1 \\ 0 & 0 & 0 & 0 & 0 \end{pmatrix}$$

$$C = \begin{pmatrix} 1 & 1 & 1 & 1 & 1 \\ 0 & 0 & 0 & 1 & 1 \end{pmatrix} \quad D = \begin{pmatrix} 2 & 0 & 1 \\ 0 & 1 & 0 \\ 0 & 0 & 1 \end{pmatrix}$$

$$E = \begin{pmatrix} 1 & 0 & 0 & 4 & 0 \\ 0 & 1 & 1 & 3 & 0 \\ 0 & 0 & 0 & 0 & 1 \\ 0 & 0 & 0 & 0 & 0 \end{pmatrix} \quad F = \begin{pmatrix} 1 & 0 & 0 & 2 & 0 \\ 0 & 0 & 2 & 1 & 0 \end{pmatrix}$$

2.4 Transpose of a matrix

Let $A = (a_{ij})$ be an $m \times n$ matrix defined in (2.1). Then the transpose of A, denoted by A^T, is an $n \times m$ matrix obtained by interchanging the rows and columns of A. Hence, $A^T = (a_{ji})$, that is, the (j, i)-entry of A^T is the (i, j)-entry of A. Hence,

$$A^T = \begin{pmatrix} a_{11} & a_{21} & \cdots & a_{m1} \\ a_{12} & a_{22} & \cdots & a_{m2} \\ \vdots & \vdots & \ddots & \vdots \\ a_{1n} & a_{2n} & \cdots & a_{mn} \end{pmatrix}. \tag{2.8}$$

From (2.8), we see that the ith column of A^T and the ith row of A are same.

Example 2.4.1. Find the transposes of the following matrices:

$$A = \begin{pmatrix} 7 & 9 \\ 18 & 31 \\ 52 & 68 \end{pmatrix}, B = \begin{pmatrix} 1 & 3 & 5 \end{pmatrix}, C = (4) \text{ and } D = \begin{pmatrix} 1 & 3 & -5 \\ -2 & 7 & 8 \\ 4 & 0 & 6 \end{pmatrix}.$$

Solution. $A^T = \begin{pmatrix} 7 & 18 & 52 \\ 9 & 31 & 68 \end{pmatrix}$, $B^T = \begin{pmatrix} 1 \\ 3 \\ 5 \end{pmatrix}$, $C^T = (4)$ and

$$D^T = \begin{pmatrix} 1 & -2 & 4 \\ 3 & 7 & 0 \\ -5 & 8 & 6 \end{pmatrix}.$$

□

The following theorem gives some properties of transpose matrices. The proofs are left to reader.

Theorem 2.4.2. *Let A be an $m \times n$ matrix. Then the following assertions hold:*

(i) $(A^T)^T = A$.

(ii) If A is symmetric, then $A^T = A$ and A^T is symmetric.

(iii) If A is an upper triangular (or a lower triangular) matrix, then A^T is a lower triangular (or an upper triangular) matrix.

Exercises

1. Find the transposes of the following matrices:

$$A = \begin{pmatrix} 33 \\ 8 \\ 12 \end{pmatrix} \quad B = \begin{pmatrix} 3 & 11 & 2 \end{pmatrix} \quad C = (4) \quad D = \begin{pmatrix} 3 & 9 & -19 \\ -2 & 8 & -7 \\ 5 & 3 & -9 \end{pmatrix}$$

2.5 Equality of matrices

Definition 2.5.1. Two matrices A and B are said to be equal if the following conditions hold:

(1) A and B have same size and

(2) All of their corresponding entries are same.

If A and B are equal, then we write $A = B$. Let $A = (a_{ij})$ and $B = (b_{ij})$ are two $m \times n$ matrices. Then $A = B$ if and only if

$$a_{ij} = b_{ij} \quad \text{for all } i \in \mathcal{I}_m \text{ and } j \in \mathcal{I}_n.$$

Example 2.5.2. Let $A = \begin{pmatrix} 2 & 1 \\ 3 & x^2 \end{pmatrix}$ and $B = \begin{pmatrix} 2 & 1 \\ 3 & 4 \end{pmatrix}$. Find all $x \in \mathbb{R}$ such that $A = B$.

Solution. Note that A and B have same size. Hence, if $x^2 = 4$, then $A = B$. Hence, when $x = 2$ or $x = -2$, $A = B$. $\square$

Example 2.5.3. Let $A = \begin{pmatrix} a & 2x+y \\ b & 4x+3y \end{pmatrix}$ and $B = \begin{pmatrix} 1 & 1 \\ 3 & 6 \end{pmatrix}$. Find all $a, b, x, y \in \mathbb{R}$ such that $A = B$.

Solution. Note that A and B have same size. Hence, $A = B$ if $a = 1$, $b = 3$ and x, y is the solution of the following system of equations:

$$\begin{cases} 2x + y = 1 \\ 4x + 3y = 6. \end{cases}$$

Solving the above system implies that $x = -3/2$ and $y = 4$ is the solution of the system. Hence, if $a = 1$, $b = 3$, $x = -3/2$ and $y = 4$, then $A = B$. $\square$

Example 2.5.4. Let $A = (1 \quad 2)$ and $B = \begin{pmatrix} 1 \\ 2 \end{pmatrix}$ be two matrices. Determine if $A = B$.

Solution. Since A and B have different sizes, $A \neq B$. $\square$

Note that in Example 2.5.4, if we treat A and B as vectors, then they are equal vectors.

Exercises

1. Let $A = \begin{pmatrix} 9 & 3 \\ 6 & x^2 \end{pmatrix}$ and $B = \begin{pmatrix} 9 & 3 \\ 6 & 4 \end{pmatrix}$. Find all $x \in \mathbb{R}$ such that $A = B$.

2. Let $C = \begin{pmatrix} 12 & 5 & 25 \\ 19 & 4 & 6 \end{pmatrix}$ and $D = \begin{pmatrix} 12 & 5 & x^2 \\ 19 & 4 & 6 \end{pmatrix}$. Find all $x \in \mathbb{R}$ such that $C = D$.

3. Let $E = \begin{pmatrix} 120 & 25 & 122 \\ 123 & 124 & 125 \\ 126 & 127 & 128 \end{pmatrix}$ and $F = \begin{pmatrix} 120 & x^2 & 122 \\ 123 & 124 & x^3 \\ 26x-4 & 127 & 128 \end{pmatrix}$. Find all $x \in \mathbb{R}$ such that $E = F$.

4. Let $A = \begin{pmatrix} a & b \\ 3x+2y & -x+y \end{pmatrix}$ and $B = \begin{pmatrix} 1 & 1 \\ 3 & 6 \end{pmatrix}$. Find all $a, b, x, y \in \mathbb{R}$ such that $A = B$.

5. Let
$$C = \begin{pmatrix} a+b & 2b-a \\ x-2y & 5x+3y \end{pmatrix} \quad \text{and } D = \begin{pmatrix} 6 & 0 \\ 8 & 14 \end{pmatrix}.$$
Find all $a, b, x, y \in \mathbb{R}$ such that $C = D$.

2.6 Operations of matrices

Definition 2.6.1. Let $A = (a_{ij})$, $B = (b_{ij})$ be $m \times n$ matrices and k a real number. Then we define

Addition: $A + B = \begin{pmatrix} a_{11}+b_{11} & a_{12}+b_{12} & \cdots & a_{1n}+b_{1n} \\ a_{21}+b_{21} & a_{22}+b_{22} & \cdots & a_{2n}+b_{2n} \\ \vdots & \vdots & \ddots & \vdots \\ a_{m1}+b_{m1} & a_{m2}+b_{m2} & \cdots & a_{mn}+b_{mn} \end{pmatrix}$.

Subtraction: $A - B = \begin{pmatrix} a_{11}-b_{11} & a_{12}-b_{12} & \cdots & a_{1n}-b_{1n} \\ a_{21}-b_{21} & a_{22}-b_{22} & \cdots & a_{2n}-b_{2n} \\ \vdots & \vdots & \ddots & \vdots \\ a_{m1}-b_{m1} & a_{m2}-b_{m2} & \cdots & a_{mn}-b_{mn} \end{pmatrix}$.

Scalar multiplication: $kA = \begin{pmatrix} ka_{11} & ka_{12} & \cdots & ka_{1n} \\ ka_{21} & ka_{22} & \cdots & ka_{2n} \\ \vdots & \vdots & \ddots & \vdots \\ ka_{m1} & ka_{m2} & \cdots & ka_{mn} \end{pmatrix}$.

Example 2.6.2. Let $A = \begin{pmatrix} 3 & 4 & 2 \\ 2 & -3 & 0 \end{pmatrix}$ and $B = \begin{pmatrix} -4 & 1 & 0 \\ 5 & -6 & 1 \end{pmatrix}$. Compute (i) $A + B$; (ii) $A - B$; (iii) $-A$; (iv) $3A$.

Solution. (i) $A + B = \begin{pmatrix} 3-4 & 4+1 & 2+0 \\ 2+5 & -3-6 & 0+1 \end{pmatrix} = \begin{pmatrix} -1 & 5 & 2 \\ 7 & -9 & 1 \end{pmatrix}$.

(ii) $A - B = \begin{pmatrix} 3-(-4) & 4-1 & 2-0 \\ 2-5 & -3-(-6) & 0-1 \end{pmatrix} = \begin{pmatrix} 7 & 3 & 2 \\ -3 & 3 & -1 \end{pmatrix}$.

(iii) $-A = \begin{pmatrix} (-1)(3) & (-1)(4) & (-1)(2) \\ (-1)(2) & (-1)(-3) & (-1)(0) \end{pmatrix} = \begin{pmatrix} -3 & -4 & -2 \\ -2 & 3 & 0 \end{pmatrix}$.

(iv) $3A = \begin{pmatrix} (3)(3) & (3)(4) & (3)(2) \\ (3)(2) & (3)(-3) & (3)(0) \end{pmatrix} = \begin{pmatrix} 9 & 12 & 6 \\ 6 & -9 & 0 \end{pmatrix}$. □

The following result can be easily proved and its proof is left to reader.

Theorem 2.6.3. *Let A, B and C are $m \times n$ matrices and let $\alpha, \beta \in \mathbb{R}$. Then*

(1) $A + 0 = A$, (2) $0A = 0$, (3) $A + B = B + A$,
(4) $(A + B) + C = A + (B + C)$, (5) $\alpha(A + B) = \alpha A + \alpha B$,
(6) $(\alpha + \beta)A = \alpha A + \beta A$, (7) $(\alpha\beta)A = \alpha(\beta A)$, (8) $1A = A$,
(9) $(A+B)^T = A^T + B^T$, (10) $(A-B)^T = A^T - B^T$, (11) $(\alpha A)^T = \alpha A^T$.

Example 2.6.4. Let

$$A = \begin{pmatrix} 1 & 0 & 1 \\ 2 & 1 & 1 \\ 0 & 0 & 2 \end{pmatrix}, B = \begin{pmatrix} 1 & 2 & 3 \\ 0 & 0 & 0 \\ 1 & 4 & 2 \end{pmatrix} \text{ and } C = \begin{pmatrix} 1 & 0 & 0 \\ 0 & 1 & 1 \\ 0 & 0 & 2 \end{pmatrix}.$$

Compute (1) $2A-B+C$; (2) $[2(A+B)]^T$; (3) $(2A)^T$.

Solution.

$$\begin{aligned}
2A-B+C &= (2A-B)+C \\
&= \left[2\begin{pmatrix} 1 & 0 & 1 \\ 2 & 1 & 1 \\ 0 & 0 & 2 \end{pmatrix} - \begin{pmatrix} 1 & 2 & 3 \\ 0 & 0 & 0 \\ 1 & 4 & 2 \end{pmatrix}\right] + \begin{pmatrix} 1 & 0 & 0 \\ 0 & 1 & 1 \\ 0 & 0 & 2 \end{pmatrix} \\
&= \left[\begin{pmatrix} 2 & 0 & 2 \\ 4 & 2 & 2 \\ 0 & 0 & 4 \end{pmatrix} - \begin{pmatrix} 1 & 2 & 3 \\ 0 & 0 & 0 \\ 1 & 4 & 2 \end{pmatrix}\right] + \begin{pmatrix} 1 & 0 & 0 \\ 0 & 1 & 1 \\ 0 & 0 & 2 \end{pmatrix} \\
&= \begin{pmatrix} 1 & -2 & -1 \\ 4 & 2 & 2 \\ -1 & -4 & 2 \end{pmatrix} + \begin{pmatrix} 1 & 0 & 0 \\ 0 & 1 & 1 \\ 0 & 0 & 2 \end{pmatrix} \\
&= \begin{pmatrix} 2 & -2 & -1 \\ 4 & 3 & 3 \\ -1 & -4 & 4 \end{pmatrix}.
\end{aligned}$$

(2) Since $A^T = \begin{pmatrix} 1 & 2 & 0 \\ 0 & 1 & 0 \\ 1 & 1 & 2 \end{pmatrix}$ and $B^T = \begin{pmatrix} 1 & 0 & 1 \\ 2 & 0 & 4 \\ 3 & 0 & 2 \end{pmatrix}$, we have

$$\begin{aligned}
[2(A+B)]^T &= 2(A+B)^T = 2(A^T+B^T) \\
&= 2\left[\begin{pmatrix} 1 & 2 & 0 \\ 0 & 1 & 0 \\ 1 & 1 & 2 \end{pmatrix} + \begin{pmatrix} 1 & 0 & 1 \\ 2 & 0 & 4 \\ 3 & 0 & 2 \end{pmatrix}\right] \\
&= 2\begin{pmatrix} 2 & 2 & 1 \\ 2 & 1 & 4 \\ 4 & 1 & 4 \end{pmatrix} = \begin{pmatrix} 4 & 4 & 2 \\ 4 & 2 & 8 \\ 8 & 2 & 8 \end{pmatrix}.
\end{aligned}$$

□

Example 2.6.5. Find the matrix A if

$$\left[2A^T - \begin{pmatrix} 1 & 0 \\ 2 & -1 \end{pmatrix}^T\right]^T = \begin{pmatrix} 0 & 1 \\ 0 & -1 \end{pmatrix}.$$

Solution. Since

$$\begin{aligned}
\left[2A^T - \begin{pmatrix} 1 & 0 \\ 2 & -1 \end{pmatrix}^T\right]^T &= (2A^T)^T - \left[\begin{pmatrix} 1 & 0 \\ 2 & -1 \end{pmatrix}^T\right]^T \\
&= 2(A^T)^T - \begin{pmatrix} 1 & 0 \\ 2 & -1 \end{pmatrix} = 2A - \begin{pmatrix} 1 & 0 \\ 2 & -1 \end{pmatrix},
\end{aligned}$$

we obtain

$$2A - \begin{pmatrix} 1 & 0 \\ 2 & -1 \end{pmatrix} = \begin{pmatrix} 0 & 1 \\ 0 & -1 \end{pmatrix}$$

and

$$2A = \begin{pmatrix} 0 & 1 \\ 0 & -1 \end{pmatrix} + \begin{pmatrix} 1 & 0 \\ 2 & -1 \end{pmatrix} = \begin{pmatrix} 1 & 1 \\ 2 & -2 \end{pmatrix}.$$

Hence, $A = \frac{1}{2}\begin{pmatrix} 1 & 1 \\ 2 & -2 \end{pmatrix} = \begin{pmatrix} \frac{1}{2} & \frac{1}{2} \\ 1 & -1 \end{pmatrix}$. □

Exercises

1. Let $A = \begin{pmatrix} -2 & 3 & 4 \\ 6 & -1 & -8 \end{pmatrix}$ and $B = \begin{pmatrix} 7 & -8 & 9 \\ 0 & -1 & 0 \end{pmatrix}$. Compute
   (i) $A+B$; (ii) $-A$; (iii) $4A-2B$; (iv) $100A+B$.

2. Let $A = \begin{pmatrix} 9 & 5 & 1 \\ 8 & 0 & 0 \\ 0 & 3 & 2 \end{pmatrix}$, $B = \begin{pmatrix} 2 & 2 & 2 \\ 0 & 0 & 0 \\ 4 & 6 & 8 \end{pmatrix}$ and $C = \begin{pmatrix} 4 & 7 & 0 \\ 0 & 3 & 1 \\ 0 & 0 & 2 \end{pmatrix}$.
   Compute (1) $3A - 2B + C$; (2) $[3(A+B)]^T$; (3) $(4A + \frac{1}{2}B - 3C)^T$.

3. Find the matrix A if $\left[(3A^T) - \begin{pmatrix} -7 & -2 \\ -6 & 9 \end{pmatrix}^T\right]^T = \begin{pmatrix} -5 & -10 \\ 33 & 12 \end{pmatrix}$.

4. Find the matrix B if

$$\left[\frac{1}{2}B + \begin{pmatrix} 6 & 3 \\ 8 & 3 \\ 1 & 4 \end{pmatrix}\right]^T - 3\begin{pmatrix} -5 & 6 \\ 8 & -9 \\ -4 & 2 \end{pmatrix}^T = \begin{pmatrix} 23 & -16 & 17 \\ -16 & 26.5 & 2 \end{pmatrix}.$$

2.7 Product of a matrix and a vector

Let

$$A = \begin{pmatrix} a_{11} & a_{12} & \cdots & a_{1n} \\ a_{21} & a_{22} & \cdots & a_{2n} \\ \vdots & \vdots & \ddots & \vdots \\ a_{m1} & a_{m2} & \cdots & a_{mn} \end{pmatrix} \quad \text{and } \overrightarrow{X} = \begin{pmatrix} x_1 \\ x_2 \\ \vdots \\ x_n \end{pmatrix}. \tag{2.9}$$

Let $\overrightarrow{r_1}, \overrightarrow{r_2}, \cdots, \overrightarrow{r_m}$ be row vectors of A,

$$\begin{aligned} \overrightarrow{r_1} &= (a_{11}, a_{12}, \cdots, a_{1n}), \\ \overrightarrow{r_2} &= (a_{21}, a_{22}, \cdots, a_{2n}), \\ &\vdots \\ \overrightarrow{r_m} &= (a_{m1}, a_{m2}, \cdots, a_{mn}). \end{aligned}$$

Then

$$\begin{cases} \overrightarrow{r_1} \cdot \overrightarrow{X} = a_{11}x_1 + a_{12}x_2 + \cdots + a_{1n}x_n \\ \overrightarrow{r_2} \cdot \overrightarrow{X} = a_{21}x_1 + a_{22}x_2 + \cdots + a_{2n}x_n \\ \vdots \\ \overrightarrow{r_m} \cdot \overrightarrow{X} = a_{m1}x_1 + a_{m2}x_2 + \cdots + a_{mn}x_n. \end{cases} \tag{2.10}$$

We define the product of A and $\overrightarrow{X}$ by

$$A\overrightarrow{X} = \begin{pmatrix} \overrightarrow{r_1} \cdot \overrightarrow{X} \\ \overrightarrow{r_2} \cdot \overrightarrow{X} \\ \vdots \\ \overrightarrow{r_m} \cdot \overrightarrow{X} \end{pmatrix},$$

that is,

$$\begin{pmatrix} a_{11} & a_{12} & \cdots & a_{1n} \\ a_{21} & a_{22} & \cdots & a_{2n} \\ \vdots & \vdots & \ddots & \vdots \\ a_{m1} & a_{m2} & \cdots & a_{mn} \end{pmatrix} \begin{pmatrix} x_1 \\ x_2 \\ \vdots \\ x_n \end{pmatrix} = \begin{pmatrix} a_{11}x_1 + a_{12}x_2 + \cdots + a_{1n}x_n \\ a_{21}x_1 + a_{22}x_2 + \cdots + a_{2n}x_n \\ \vdots \\ a_{m1}x_1 + a_{m2}x_2 + \cdots + a_{mn}x_n \end{pmatrix}.$$

We note that $A\overrightarrow{X}$ is a vector in $\mathbb{R}^m$, where m is the number of rows of A while $\overrightarrow{X}$ is a vector in $\mathbb{R}^n$, where n is the number of columns of A.

Example 2.7.1. Let $A = \begin{pmatrix} 1 & 2 & 0 \\ 3 & -1 & 4 \end{pmatrix}$, $\overrightarrow{X} = \begin{pmatrix} x_1 \\ x_2 \\ x_3 \end{pmatrix}$, $\overrightarrow{X_1} = \begin{pmatrix} 1 \\ 2 \\ 1 \end{pmatrix}$ and $\overrightarrow{X_2} = \begin{pmatrix} 0 \\ 2 \\ 2 \end{pmatrix}$. Compute $A\overrightarrow{X}$, $A\overrightarrow{X_1}$ and $A\overrightarrow{X_2}$.

Solution.

$$\begin{aligned} A\overrightarrow{X} &= \begin{pmatrix} 1 & 2 & 0 \\ 3 & -1 & 4 \end{pmatrix} \begin{pmatrix} x_1 \\ x_2 \\ x_3 \end{pmatrix} = \begin{pmatrix} (1)(x_1) + (2)(x_2) + (0)(x_3) \\ (3)(x_1) + (-1)(x_2) + (4)(x_3) \end{pmatrix} \\ &= \begin{pmatrix} x_1 + 2x_2 \\ 3x_1 - x_2 + 4x_3 \end{pmatrix}. \end{aligned}$$

$$A\overrightarrow{X_1} = \begin{pmatrix} 1 & 2 & 0 \\ 3 & -1 & 4 \end{pmatrix} \begin{pmatrix} 1 \\ 2 \\ 1 \end{pmatrix} = \begin{pmatrix} (1)(1) + (2)(2) + (0)(1) \\ (3)(1) + (-1)(2) + (4)(1) \end{pmatrix} = \begin{pmatrix} 5 \\ 5 \end{pmatrix}.$$

$$A\overrightarrow{X_2} = \begin{pmatrix} 1 & 2 & 0 \\ 3 & -1 & 4 \end{pmatrix}\begin{pmatrix} 0 \\ 2 \\ 2 \end{pmatrix} = \begin{pmatrix} (1)(0)+(2)(2)+(0)(2) \\ (3)(0)+(-1)(2)+(4)(2) \end{pmatrix} = \begin{pmatrix} 4 \\ 6 \end{pmatrix}.$$

□

Example 2.7.2. $A = \begin{pmatrix} -4 & 1 \\ 5 & -1 \\ 3 & 0 \end{pmatrix}$, $\overrightarrow{X} = \begin{pmatrix} 1 \\ 1 \end{pmatrix}$ and $\overrightarrow{X_1} = \begin{pmatrix} 0 \\ 4 \end{pmatrix}$.

Compute $A\overrightarrow{X}$ and $A\overrightarrow{X_1}$,

Solution.

$$A\overrightarrow{X} = \begin{pmatrix} -4 & 1 \\ 5 & -1 \\ 3 & 0 \\ 2 & 4 \end{pmatrix}\begin{pmatrix} 1 \\ 1 \end{pmatrix} = \begin{pmatrix} (-4)(1)+(1)(1) \\ (5)(1)+(-1)(1) \\ (3)(1)+(0)(1) \\ (2)(1)+(4)(1) \end{pmatrix} = \begin{pmatrix} -3 \\ 4 \\ 3 \\ 6 \end{pmatrix}.$$

$$A\overrightarrow{X_1} = \begin{pmatrix} -4 & 1 \\ 5 & -1 \\ 3 & 0 \\ 2 & 4 \end{pmatrix}\begin{pmatrix} 0 \\ 4 \end{pmatrix} = \begin{pmatrix} -4(0)+(1)(4) \\ (5)(0)+(-1)(4) \\ (3)(0)+(0)(4) \\ (2)(0)+(4)(4) \end{pmatrix} = \begin{pmatrix} 4 \\ -4 \\ 0 \\ 16 \end{pmatrix}.$$

□

From (2.10), we see that a matrix and a vector can be multiplied together only when the number of columns of the matrix equals the number of components of the vector. Hence, if the vectors $\overrightarrow{r_i}$ and $\overrightarrow{X}$ have different numbers of components, then the scalar product $\overrightarrow{r_i} \cdot \overrightarrow{X}$ is not defined.

Example 2.7.3. Let $A = \begin{pmatrix} -1 & 1 \\ 2 & -1 \\ 3 & 1 \end{pmatrix}$, $\overrightarrow{X_1} = \begin{pmatrix} 1 \\ 1 \\ -1 \end{pmatrix}$, $\overrightarrow{X_2} = \begin{pmatrix} 0 \\ 0 \end{pmatrix}$ and $\overrightarrow{X_3} = \begin{pmatrix} 1 \\ 1 \\ 2 \\ 4 \end{pmatrix}$. Determine whether $A\overrightarrow{X_i}$ is defined for each $i = 1, 2, 3$.

Solution. $A\overrightarrow{X_2}$ is defined, but $A\overrightarrow{X_1}$ and $A\overrightarrow{X_3}$ are not defined. □

Theorem 2.7.4. *Let A be same as in (2.9), $\overrightarrow{X}, \overrightarrow{Y} \in \mathbb{R}^n$ and $\alpha, \beta \in \mathbb{R}$. Then*

$$A(\alpha\overrightarrow{X} + \beta\overrightarrow{Y}) = \alpha(A\overrightarrow{X}) + \beta(A\overrightarrow{Y}).$$

Example 2.7.5. Let

$$A = \begin{pmatrix} 1 & 1 & 0 \\ 2 & 0 & -1 \\ 1 & 0 & 1 \end{pmatrix}, \quad \overrightarrow{X} = \begin{pmatrix} 1 \\ 2 \\ -1 \end{pmatrix} \quad \text{and } \overrightarrow{Y} = \begin{pmatrix} 0 \\ 1 \\ -1 \end{pmatrix}.$$

Compute $A(2\overrightarrow{X} + 3\overrightarrow{Y})$.

Solution. Let $\overrightarrow{a} = A(2\overrightarrow{X} + 3\overrightarrow{Y})$. Then

$$\begin{aligned}
\overrightarrow{a} &= A(2\overrightarrow{X} + 3\overrightarrow{Y}) = 2(A\overrightarrow{X}) + 3(A\overrightarrow{Y}) \\
&= 2\begin{pmatrix} 1 & 1 & 0 \\ 2 & 0 & -1 \\ 1 & 0 & 1 \end{pmatrix}\begin{pmatrix} 1 \\ 2 \\ -1 \end{pmatrix} + 3\begin{pmatrix} 1 & 1 & 0 \\ 2 & 0 & -1 \\ 1 & 0 & 1 \end{pmatrix}\begin{pmatrix} 0 \\ 1 \\ -1 \end{pmatrix} \\
&= 2\begin{pmatrix} 3 \\ 3 \\ 0 \end{pmatrix} + 3\begin{pmatrix} 1 \\ 1 \\ -1 \end{pmatrix} = \begin{pmatrix} 6 \\ 6 \\ 0 \end{pmatrix} + \begin{pmatrix} 3 \\ 3 \\ -3 \end{pmatrix} = \begin{pmatrix} 9 \\ 9 \\ -3 \end{pmatrix}.
\end{aligned}$$

□

Let $\overrightarrow{a_1}, \overrightarrow{a_2}, \cdots, \overrightarrow{a_n}$ be column vectors of A given in (2.2). Then $A\overrightarrow{X}$ can be expressed by a linear combination of these column vectors of A:

$$A\overrightarrow{X} = x_1\overrightarrow{a_1} + x_2\overrightarrow{a_2} + \cdots + x_n\overrightarrow{a_n}. \tag{2.11}$$

Example 2.7.6. Let $A = \begin{pmatrix} 1 & 2 & 3 \\ 4 & 5 & 6 \\ 7 & 8 & 9 \end{pmatrix}$ and $\overrightarrow{X} = \begin{pmatrix} 2 \\ 3 \\ 4 \end{pmatrix}$. Write $A\overrightarrow{X}$ as a linear combination of the column vectors of A.

Solution. $A\overrightarrow{X} = 2\begin{pmatrix} 1 \\ 4 \\ 7 \end{pmatrix} + 3\begin{pmatrix} 2 \\ 5 \\ 8 \end{pmatrix} + 4\begin{pmatrix} 3 \\ 6 \\ 9 \end{pmatrix}$.

Exercises

1. Let $A = \begin{pmatrix} 2 & 1 & 1 \\ 1 & -1 & 2 \end{pmatrix}$, $\overrightarrow{X} = \begin{pmatrix} x_1 \\ x_2 \\ x_3 \end{pmatrix}$, $\overrightarrow{X_1} = \begin{pmatrix} 0 \\ 1 \\ 2 \end{pmatrix}$ and $\overrightarrow{X_2} = \begin{pmatrix} 2 \\ 1 \\ 1 \end{pmatrix}$. Compute $A\overrightarrow{X}$, $A\overrightarrow{X_1}$ and $A\overrightarrow{X_2}$.

2. $A = \begin{pmatrix} -2 & -1 \\ 3 & 1 \\ 0 & 1 \end{pmatrix}$, $\overrightarrow{X} = \begin{pmatrix} -a \\ 2a \end{pmatrix}$ and $\overrightarrow{X_1} = \begin{pmatrix} 1 \\ 2 \end{pmatrix}$.

   Compute $A\overrightarrow{X}$ and $A\overrightarrow{X_1}$.

3. Let $A = \begin{pmatrix} 0 & 1 \\ 1 & -1 \\ 2 & 1 \end{pmatrix}$, $\overrightarrow{X_1} = \begin{pmatrix} 1 \\ 0 \\ -1 \end{pmatrix}$, $\overrightarrow{X_2} = \begin{pmatrix} 1 \\ 0 \end{pmatrix}$ and $\overrightarrow{X_3} = \begin{pmatrix} 1 \\ 1 \\ 1 \\ 2 \end{pmatrix}$.

   Determine whether $A\overrightarrow{X_i}$ is defined for each $i = 1, 2, 3$.

4. Let $A = \begin{pmatrix} 0 & 1 & 0 \\ 1 & 0 & -1 \\ 1 & 1 & 2 \end{pmatrix}$, $\overrightarrow{X} = \begin{pmatrix} 1 \\ 0 \\ -1 \end{pmatrix}$ and $\overrightarrow{Y} = \begin{pmatrix} 3 \\ 0 \\ -1 \end{pmatrix}$.

   Compute $A(\overrightarrow{X} - 3\overrightarrow{Y})$.

5. Let $A = \begin{pmatrix} 1 & 1 & -1 \\ 1 & 0 & 2 \\ 2 & 1 & -1 \end{pmatrix}$ and $\overrightarrow{X} = \begin{pmatrix} 1 \\ -1 \\ 0 \end{pmatrix}$.

   Write $A\overrightarrow{X}$ as a linear combination of the column vectors of A.

2.8 Product of two matrices

Let

$$A = \begin{pmatrix} a_{11} & a_{12} & \cdots & a_{1n} \\ a_{21} & a_{22} & \cdots & a_{2n} \\ \vdots & \vdots & \ddots & \vdots \\ a_{m1} & a_{m2} & \cdots & a_{mn} \end{pmatrix} \quad \text{and } B_{n\times r} = \begin{pmatrix} b_{11} & b_{12} & \cdots & b_{1r} \\ b_{21} & b_{22} & \cdots & b_{2r} \\ \cdots & \cdots & \ddots & \cdots \\ b_{n1} & b_{n2} & \cdots & b_{nr} \end{pmatrix}.$$

Let $\overrightarrow{r_1}, \overrightarrow{r_2}, \cdots, \overrightarrow{r_m}$ be row vectors of A, that is,

$$\begin{aligned} \overrightarrow{r_1} &= (a_{11}, a_{12}, \cdots, a_{1n}), \\ \overrightarrow{r_2} &= (a_{21}, a_{22}, \cdots, a_{2n}), \\ &\vdots \\ \overrightarrow{r_m} &= (a_{m1}, a_{m2}, \cdots, a_{mn}). \end{aligned}$$

Let $\overrightarrow{b_1} = \begin{pmatrix} b_{11} \\ b_{21} \\ \vdots \\ b_{n1} \end{pmatrix}$, $\overrightarrow{b_2} = \begin{pmatrix} b_{12} \\ b_{22} \\ \vdots \\ b_{n2} \end{pmatrix}$, $\cdots$, $\overrightarrow{b_r} = \begin{pmatrix} b_{1r} \\ b_{2r} \\ \vdots \\ b_{nr} \end{pmatrix}$ are column vectors of B. Then the product AB of A and B is defined by

$$AB = (A\overrightarrow{b_1}\ A\overrightarrow{b_2}\ \cdots\ A\overrightarrow{b_r}). \tag{2.12}$$

Hence,

$$AB = \begin{pmatrix} \overrightarrow{r_1}\cdot\overrightarrow{b_1} & \overrightarrow{r_1}\cdot\overrightarrow{b_2} & \cdots & \overrightarrow{r_1}\cdot\overrightarrow{b_r} \\ \overrightarrow{r_2}\cdot\overrightarrow{b_1} & \overrightarrow{r_2}\cdot\overrightarrow{b_2} & \cdots & \overrightarrow{r_2}\cdot\overrightarrow{b_r} \\ \vdots & \vdots & \ddots & \vdots \\ \overrightarrow{r_m}\cdot\overrightarrow{b_1} & \overrightarrow{r_m}\cdot\overrightarrow{b_2} & \cdots & \overrightarrow{r_m}\cdot\overrightarrow{b_r} \end{pmatrix}. \tag{2.13}$$

Moreover, the size of AB is $m \times r$.

Example 2.8.1. Let $A = \begin{pmatrix} 1 & 3 \\ -2 & 4 \end{pmatrix}$ and $B = \begin{pmatrix} 3 & -2 & 0 \\ 5 & 6 & 0 \end{pmatrix}$. Find AB and the sizes of A, B and AB.

Solution.

$$\begin{aligned} AB &= \begin{pmatrix} (1)(3)+(3)(5) & (1)(-2)+(3)(6) & (1)(0)+(3)(0) \\ (-2)(3)+(4)(5) & (-2)(-2)+(4)(6) & (-2)(0)+(4)(0) \end{pmatrix} \\ &= \begin{pmatrix} 18 & 16 & 0 \\ 14 & 28 & 0 \end{pmatrix}. \end{aligned}$$

The sizes of A, B and AB are 2×2, 2×3, 2×3, respectively. □

Example 2.8.2. Let $A = \begin{pmatrix} 1 & 2 & 4 \\ 2 & 6 & 0 \end{pmatrix}$ and $B = \begin{pmatrix} 4 & 1 & 4 \\ 0 & -1 & 3 \\ 2 & 7 & 5 \end{pmatrix}$. Compute AB and find the sizes of A, B and AB.

Solution. By computation, we have

$$AB = \begin{pmatrix} 1 & 2 & 4 \\ 2 & 6 & 0 \end{pmatrix} \begin{pmatrix} 4 & 1 & 4 \\ 0 & -1 & 3 \\ 2 & 7 & 5 \end{pmatrix} = \begin{pmatrix} 12 & 27 & 30 \\ 8 & -4 & 26 \end{pmatrix}.$$

The sizes of A, B and AB are 2×3, 3×3, 2×3, respectively. □

Example 2.8.3. Let $A = \begin{pmatrix} 2 & 0 & -3 \\ 4 & 1 & 5 \end{pmatrix}$ and $B = \begin{pmatrix} 7 & -1 & 4 & 7 \\ 2 & 5 & 0 & -4 \\ -3 & 1 & 2 & 3 \end{pmatrix}$. Compute AB and find the sizes of A, B and AB.

Solution.

$$AB = \begin{pmatrix} 2 & 0 & -3 \\ 4 & 1 & 5 \end{pmatrix} \begin{pmatrix} 7 & -1 & 4 & 7 \\ 2 & 5 & 0 & -4 \\ -3 & 1 & 2 & 3 \end{pmatrix} = \begin{pmatrix} 23 & -5 & 2 & 5 \\ 15 & 6 & 26 & 39 \end{pmatrix}.$$

The sizes of A, B and AB are 2×3, 3×3, 2×3, respectively. □

Exercises

1. Let $A = \begin{pmatrix} 2 & 1 \\ -2 & 3 \end{pmatrix}$ and $B = \begin{pmatrix} 1 & -2 & 1 \\ 3 & 4 & 1 \end{pmatrix}$.

   Find AB and the sizes of A, B and AB.

2. Let $A = \begin{pmatrix} 0 & 1 & -1 \\ 2 & 3 & 1 \end{pmatrix}$ and $B = \begin{pmatrix} 2 & 1 & -1 \\ 1 & 0 & 1 \\ -1 & 2 & 4 \end{pmatrix}$.

   Compute AB and find the sizes of A, B and AB.

3. Let $A = \begin{pmatrix} 1 & 0 & -2 \\ 3 & 2 & -1 \end{pmatrix}$ and $B = \begin{pmatrix} 0 & -1 & 2 & 1 \\ -1 & 2 & 1 & -2 \\ 2 & 0 & 0 & 1 \end{pmatrix}$.

   Compute AB and find the sizes of A, B and AB.

2.9 Sizes of two matrices in a product

We have known that a matrix and a vector can be multiplied together only when the number of columns of the matrix equals the number of components of the vector.

In (2.12), for each $i = 1, 2, \cdots, r$, the product $A\overrightarrow{b_i}$ requires that the number of columns of the matrix A is equal to the number of components of the vector $\overrightarrow{b_i}$. Hence, the product of two matrices A and B requires that the number of columns of A is equal to the number of rows of B. Symbolically,

$$(m \times n)(n \times r) = m \times r.$$

The inner numbers must be same and outer numbers give the size of the product AB. Hence, if the inner numbers are not same, the product AB is not defined.

Example 2.9.1. Let $A = \begin{pmatrix} 1 & 0 \\ 2 & 1 \\ 3 & 0 \end{pmatrix}$ and $B = \begin{pmatrix} 1 & 2 \\ 3 & 4 \end{pmatrix}$. Are AB and BA defined? If so, compute them. If not, explain why?

Solution. The size of A is 3×2 and the size of B is 2×2. Hence, AB is defined since the number of columns of A and the number of row of B are same and equal 2, but BA is not defined since the number of columns of B is 2 and the number of row of A is 3 and they are not same.

$$AB = \begin{pmatrix} 1 & 0 \\ 2 & 1 \\ 3 & 0 \end{pmatrix} \begin{pmatrix} 1 & 2 \\ 3 & 4 \end{pmatrix} = \begin{pmatrix} 1 & 2 \\ 5 & 8 \\ 3 & 6 \end{pmatrix}.$$

□

Example 2.9.2. Let $A = \begin{pmatrix} 1 & 3 \\ -2 & 4 \end{pmatrix}$ and $B = \begin{pmatrix} 3 & -2 \\ 5 & 6 \end{pmatrix}$. Are AB and BA defined? If so, compute them. If not, explain why? is AB equal to BA?

Solution. AB and BA are defined since their sizes are 2×2.

$$AB = \begin{pmatrix} 1 & 3 \\ -2 & 4 \end{pmatrix} \begin{pmatrix} 3 & -2 \\ 5 & 6 \end{pmatrix} = \begin{pmatrix} 18 & 16 \\ 14 & 28 \end{pmatrix}$$

and

$$BA = \begin{pmatrix} 3 & -2 \\ 5 & 6 \end{pmatrix} \begin{pmatrix} 1 & 3 \\ -2 & 4 \end{pmatrix} = \begin{pmatrix} 7 & 1 \\ -7 & 39 \end{pmatrix}.$$

Hence, $AB \neq BA$. □

Exercises

1. Let $A = \begin{pmatrix} 1 & 1 \\ 1 & 3 \\ 3 & 1 \end{pmatrix}$ and $B = \begin{pmatrix} 2 & 1 \\ -2 & 3 \end{pmatrix}$. Are AB and BA defined? If so, compute them. If not, explain why?

2. Let $A = \begin{pmatrix} 2 & -1 \\ -2 & 3 \end{pmatrix}$ and $B = \begin{pmatrix} 0 & -2 \\ 4 & 2 \end{pmatrix}$. Are AB and BA defined? If so, compute them. If not, explain why? Is AB equal to BA?

2.10 Properties of products of matrices

Theorem 2.10.1. *The following assertions hold:*

(1) *(Associative law for matrix multiplication)*
Let $A := A_{m\times n}$, $B := B_{n\times p}$, *and* $C := C_{p\times q}$. *Then*

$$A(BC) = (AB)C.$$

(2) *(Distributive laws for matrix multiplication)*
(*i*) *Let* $A := A_{m\times n}$, $B := B_{n\times p}$, *and* $C := C_{n\times q}$. *Then*

$$A(B + C) + AB + AC.$$

(*ii*) *Let* $A := A_{m\times n}$, $B := B_{m\times n}$, *and* $C := C_{n\times q}$. *Then*

$$(A + B)C = AC + BC.$$

Example 2.10.2. Let

$$A = \begin{pmatrix} 1 & -3 \\ 0 & 2 \end{pmatrix}, \ B = \begin{pmatrix} 2 & -1 & 4 \\ 3 & 1 & 5 \end{pmatrix} \quad \text{and } C = \begin{pmatrix} 0 & -2 & 1 \\ 4 & 3 & 2 \\ -5 & 0 & 6 \end{pmatrix}.$$

Show that $A(BC) = (AB)C$.

Solution.

$$BC = \begin{pmatrix} 2 & -1 & 4 \\ 3 & 1 & 5 \end{pmatrix} \begin{pmatrix} 0 & -2 & 1 \\ 4 & 3 & 2 \\ -5 & 0 & 6 \end{pmatrix} = \begin{pmatrix} -24 & -7 & 24 \\ -21 & -3 & 35 \end{pmatrix}$$

$$A(BC) = \begin{pmatrix} 1 & -3 \\ 0 & 2 \end{pmatrix} \begin{pmatrix} -24 & -7 & 24 \\ -21 & -3 & 35 \end{pmatrix} = \begin{pmatrix} 39 & 2 & -81 \\ -42 & -6 & 70 \end{pmatrix}$$

$$AB = \begin{pmatrix} 1 & -3 \\ 0 & 2 \end{pmatrix} \begin{pmatrix} 2 & -1 & 4 \\ 3 & 1 & 5 \end{pmatrix} = \begin{pmatrix} -7 & -4 & -11 \\ 6 & 2 & 10 \end{pmatrix}$$

$$(AB)C = \begin{pmatrix} -7 & -4 & -11 \\ 6 & 2 & 10 \end{pmatrix} \begin{pmatrix} 0 & -2 & 1 \\ 4 & 3 & 2 \\ -5 & 0 & 6 \end{pmatrix} = \begin{pmatrix} 39 & 2 & -81 \\ -42 & -6 & 70 \end{pmatrix}.$$

Hence, $A(BC) = (AB)C$. □

Exercises

1. Let

$$A = \begin{pmatrix} 1 & -1 \\ -1 & 2 \end{pmatrix}, \; B = \begin{pmatrix} 0 & -1 & 2 \\ 1 & 1 & 3 \end{pmatrix} \quad \text{and } C = \begin{pmatrix} 1 & -1 & 1 \\ 0 & 3 & 2 \\ -2 & 0 & 3 \end{pmatrix}.$$

Show that $A(BC) = (AB)C$.

2.11 Powers of a square matrix

Let A be a square matrix of order n given in (2.6) and I_n the identity matrix, that is,

$$A = \begin{pmatrix} a_{11} & a_{12} & \cdots & a_{1n} \\ a_{21} & a_{22} & \cdots & a_{2n} \\ \vdots & \vdots & \ddots & \vdots \\ a_{n1} & a_{n2} & \cdots & a_{nn} \end{pmatrix} \quad \text{and } I_n = \begin{pmatrix} 1 & 0 & \cdots & 0 \\ 0 & 1 & \cdots & 0 \\ \vdots & \vdots & \ddots & \vdots \\ 0 & 0 & \cdots & 1 \end{pmatrix}.$$

We define

$$A^0 = I_n, \quad A^2 = A \cdot A, \quad A^3 = A^2 \cdot A, \cdots, A^n = A^{n-1} \cdot A. \tag{2.14}$$

Example 2.11.1. Let $A = \begin{pmatrix} 1 & 2 \\ 1 & 3 \end{pmatrix}$. Find A^2 and A^3.

Solution. $A^2 = \begin{pmatrix} 1 & 2 \\ 1 & 3 \end{pmatrix}\begin{pmatrix} 1 & 2 \\ 1 & 3 \end{pmatrix} = \begin{pmatrix} 3 & 8 \\ 4 & 11 \end{pmatrix}$ and

$$A^3 = A^2 A = \begin{pmatrix} 3 & 8 \\ 4 & 11 \end{pmatrix}\begin{pmatrix} 1 & 2 \\ 1 & 3 \end{pmatrix} = \begin{pmatrix} 11 & 30 \\ 15 & 41 \end{pmatrix}.$$

□

Theorem 2.11.2. *Let* $r, s \in \mathbb{N}$. *Then the following properties hold:*
(1) $A^r \cdot A^s = A^{r+s}$.
(2) $(A^r)^s = A^{rs}$.

Let

$$P(x) = a_0 + a_1 x + \cdots + a_m x^m$$

is a polynomial. Then we define the following matrix:

$$P(A) = a_0 I_n + a_1 A + a_2 A^2 + \cdots + a_m A^m. \tag{2.15}$$

Example 2.11.3. Let $P(x) = 4 - 3x + 2x^2$ and $A = \begin{pmatrix} -1 & 2 \\ 0 & 3 \end{pmatrix}$. Compute $P(A)$.

Solution.

$$\begin{aligned} P(A) &= 4I_2 - 3A + 2A^2 = 4\begin{pmatrix} 1 & 0 \\ 0 & 1 \end{pmatrix} - 3\begin{pmatrix} -1 & 2 \\ 0 & 3 \end{pmatrix} + 2\begin{pmatrix} -1 & 2 \\ 0 & 3 \end{pmatrix}^2 \\ &= \begin{pmatrix} 4 & 0 \\ 0 & 4 \end{pmatrix} - \begin{pmatrix} -3 & 6 \\ 0 & 9 \end{pmatrix} + \begin{pmatrix} 2 & 8 \\ 0 & 18 \end{pmatrix} = \begin{pmatrix} 9 & 2 \\ 0 & 13 \end{pmatrix}. \end{aligned}$$

□

Theorem 2.11.4. (*i*) *If* A *and* B *are lower triangular matrices, then* AB *is a lower triangular matrix.*

(*ii*) *If* A *and* B *are upper triangular matrices, then* AB *is an upper triangular matrix.*

The following example shows that the product of two lower triangular matrices is a lower triangular matrix.

Example 2.11.5. $A = \begin{pmatrix} 1 & 0 \\ 2 & 1 \end{pmatrix}$ and $B = \begin{pmatrix} 3 & 0 \\ 4 & 1 \end{pmatrix}$. Compute AB.

Solution. $AB = \begin{pmatrix} 1 & 0 \\ 2 & 1 \end{pmatrix}\begin{pmatrix} 3 & 0 \\ 4 & 1 \end{pmatrix} = \begin{pmatrix} 3 & 0 \\ 10 & 1 \end{pmatrix}$. □

Let $k \in \mathbb{N}$ and let

$$A = \begin{pmatrix} a_1 & 0 & \cdots & 0 \\ 0 & a_2 & \cdots & 0 \\ \vdots & \vdots & \ddots & \vdots \\ 0 & 0 & \cdots & a_n \end{pmatrix}$$

be a diagonal matrix. Then it is easy to verify that

$$A^k = \begin{pmatrix} a_1^k & 0 & \cdots & 0 \\ 0 & a_2^k & \cdots & 0 \\ \vdots & \vdots & \ddots & \vdots \\ 0 & 0 & \cdots & a_n^k \end{pmatrix}. \tag{2.16}$$

□

Example 2.11.6. Let $A = \begin{pmatrix} 1 & 0 & 0 \\ 0 & -2 & 0 \\ 0 & 0 & 2 \end{pmatrix}$ Find A^5.

Solution. Since A is a diagonal matrix, it follows from (2.16) that

$$A^5 = \begin{pmatrix} 1^5 & 0 & 0 \\ 0 & (-2)^5 & 0 \\ 0 & 0 & 2^5 \end{pmatrix} = \begin{pmatrix} 1 & 0 & 0 \\ 0 & -32 & 0 \\ 0 & 0 & 32 \end{pmatrix}.$$

□

Exercises

1. Let $A = \begin{pmatrix} 1 & 0 \\ -1 & 1 \end{pmatrix}$. Find A^2, A^3 and A^4.

2. Let $P(x) = 1 - 2x - x^2$ and $A = \begin{pmatrix} -1 & 1 \\ 1 & 2 \end{pmatrix}$. Compute $P(A)$.

3. Let $A = \begin{pmatrix} 2 & 0 & 0 \\ 0 & -2 & 0 \\ 0 & 0 & 3 \end{pmatrix}$. Find A^6.

2.12 Row operations of matrices

In many cases, we need to change a matrix to a row echelon matrix or a reduced row echelon matrix and wish that the row echelon matrix or the reduced row echelon matrix and the original matrix share some common properties. Hence, one can obtain properties of the original matrix by studying the row echelon matrix or the reduced row echelon matrix. The powerful tool to change a matrix to a row echelon matrix or reduced row echelon matrix is to use the following (elementary) row operations.

1. **First row operation** $R_i(c)$: Multiply ith row of a matrix by a nonzero number c;

2. **Second row operation** $R_i(c) + R_j$: Add ith row multiplied by c to jth row;

3. **Third row operation** $R_{i,j}$: Interchange ith row and jth row.

Remark 2.12.1. The first row operation $R_i(c)$ is used to change ith row, so other rows keep unchanged. The second row operation $R_i(c) + R_j$ is used to change jth row, so other rows including the ith row keep unchanged. The third row operation $R_{i,j}$ is used to interchange ith row and jth row, so other rows keep unchanged.

Example 2.12.2. Let $A = \begin{pmatrix} 0 & 1 & 1 \\ 1 & 2 & -3 \\ 2 & 1 & 1 \end{pmatrix}$.

(*i*) We apply $R_2(-2)$ to change the row 2 of A, so the rows 1 and 3 keep unchanged.

$$A = \begin{pmatrix} 0 & 1 & 1 \\ 1 & 2 & -3 \\ 2 & 1 & 1 \end{pmatrix} \xrightarrow{R_2(-2)} \begin{pmatrix} 0 & 1 & 1 \\ -2 & -4 & 6 \\ 2 & 1 & 1 \end{pmatrix}.$$

(*ii*) We apply $R_2(-2) + R_3$ to change the row 3 of A, so the rows 1 and 2 keep unchanged.

$$A = \begin{pmatrix} 0 & 1 & 1 \\ 1 & 2 & -3 \\ 2 & 1 & 1 \end{pmatrix} \xrightarrow{R_2(-2) + R_3} \begin{pmatrix} 0 & 1 & 1 \\ 1 & 2 & -3 \\ 0 & -3 & 7 \end{pmatrix}.$$

(*iii*) We apply $R_{1,2}$ to change the rows 1 and 2 of A, so the row 3 keeps unchanged.

$$A = \begin{pmatrix} 0 & 1 & 1 \\ 1 & 2 & -3 \\ 2 & 1 & 1 \end{pmatrix} \xrightarrow{R_{1,2}} \begin{pmatrix} 1 & 2 & -3 \\ 0 & 1 & 1 \\ 2 & 1 & 1 \end{pmatrix}$$

In the following we shall show the process how to use row operations to find a row echelon matrix for a given matrix.

Example 2.12.3. Let $A = \begin{pmatrix} 1 & 1 & 1 \\ 2 & -1 & -1 \\ -3 & 4 & 1 \end{pmatrix}$. Use the second row operation to make the numbers 2 and -3 in A to be zero.

Solution. To change the number 2 in the first column or in the second row of A, we use the row operation $R_1(-2) + R_2$ to change the entire row 2.

$$A = \begin{pmatrix} 1 & 1 & 1 \\ 2 & -1 & -1 \\ -3 & 4 & 1 \end{pmatrix} \xrightarrow{R_1(-2)+R_2} \begin{pmatrix} 1 & 1 & 1 \\ 0 & -3 & -3 \\ -3 & 4 & 1 \end{pmatrix}.$$

Similarly, we use $R_1(3) + R_3$ to change the entire row 3 in order to change the number -3 to be 0.

$$\begin{pmatrix} 1 & 1 & 1 \\ 0 & -3 & -3 \\ -3 & 4 & 1 \end{pmatrix} \xrightarrow{R_1(3)+R_3} \begin{pmatrix} 1 & 1 & 1 \\ 0 & -3 & -3 \\ 0 & 7 & 4 \end{pmatrix}.$$

□

From the above, we see that row 1 is always kept unchanged when we change rows 2 and 3. Hence, we can combine the process as follows:

$$A = \begin{pmatrix} 1 & 1 & 1 \\ 2 & -1 & -1 \\ -3 & 4 & 1 \end{pmatrix} \xrightarrow[R_1(3)+R_3]{R_1(-2)+R_2} \begin{pmatrix} 1 & 1 & 1 \\ 0 & -3 & -3 \\ 0 & 7 & 4 \end{pmatrix}.$$

Example 2.12.4. Let $B = \begin{pmatrix} 1 & 1 & 1 \\ 0 & -3 & -3 \\ 0 & 7 & 4 \end{pmatrix}$. Use the second row operation to make the number 7 in B to be zero.

Solution. $B = \begin{pmatrix} 1 & 1 & 1 \\ 0 & -3 & -3 \\ 0 & 7 & 4 \end{pmatrix} \xrightarrow{R_2(\frac{7}{3})+R_3} \begin{pmatrix} 1 & 1 & 1 \\ 0 & -3 & -3 \\ 0 & 0 & -3 \end{pmatrix}.$ □

Now, we combine Examples 2.12.3 and 2.12.4 to change A to a row echelon matrix.

Example 2.12.5. Let A be same as in Example 2.12.3. Find a row echelon matrix of the matrix A.

Solution.

$$A = \begin{pmatrix} 1 & 1 & 1 \\ 2 & -1 & -1 \\ -3 & 4 & 1 \end{pmatrix} \xrightarrow[R_1(3)+R_3]{R_1(-2)+R_2} \begin{pmatrix} 1 & 1 & 1 \\ 0 & -3 & -3 \\ 0 & 7 & 4 \end{pmatrix}$$

$$\xrightarrow{R_2(\frac{7}{3})+R_3} \begin{pmatrix} 1 & 1 & 1 \\ 0 & -3 & -3 \\ 0 & 0 & -3 \end{pmatrix}.$$

□

After we get some ideas how to change a matrix to its echelon matrix from Examples 2.12.3 and 2.12.4, we study examples below which show the method.

Example 2.12.6. Find a row echelon matrix of

$$A = \begin{pmatrix} 0 & 2 & -1 \\ 1 & -1 & 3 \\ 2 & 0 & 1 \end{pmatrix}.$$

Solution. (1) Check the first column to see if there are nonzero numbers. If all entries in the first column are zero, go to the second column. If there are nonzero numbers in the first column, use the third row operation to make the first number (a_{11}) in the first row be nonzero.

In the matrix A, the first column contains a nonzero number 1 which is in the second row. We use $R_{1,2}$ to change A to

$$\begin{pmatrix} 1 & -1 & 3 \\ 0 & 2 & -1 \\ 2 & 0 & 1 \end{pmatrix}.$$

So, the first number of the first row of the new matrix is nonzero.

We write the process as follows:

$$A = \begin{pmatrix} 0 & 2 & -1 \\ 1 & -1 & 3 \\ 2 & 0 & 1 \end{pmatrix} \xrightarrow{R_{1,2}} \begin{pmatrix} 1 & -1 & 3 \\ 0 & 2 & -1 \\ 2 & 0 & 1 \end{pmatrix}.$$

(2) Use the second row operation to change all nonzero entries below a_{11} in the first column to be zero.

In the last matrix, we need to change the first number in the third row to be zero, so we use $R_1(-2) + R_3$ to change row 3.

$$\begin{pmatrix} 1 & -1 & 3 \\ 0 & 2 & -1 \\ 2 & 0 & 1 \end{pmatrix} \xrightarrow{R_1(-2)+R_3} \begin{pmatrix} 1 & -1 & 3 \\ 0 & 2 & -1 \\ 0 & 2 & -5 \end{pmatrix} := B.$$

(3) In the following, we shall use the same method as in (1) to treat the other rows and columns of the last matrix B. We check the second column. First, check the second column to see if the number in the position a_{22} is zero. If all entries below the first number in the second column are zero, go to the third column. If there is one nonzero number below a_{22} in the second column, use the second row operation to interchange row 2 and one of other rows except row 1 to make the number in the position a_{22} to be nonzero. Then use the second row operation to change all nonzero entries below a_{22} in the second column to be zero.

In the last matrix B, the number in the position a_{22} is nonzero. So, we use the second row operation $R_2(-1)+R_3$ to change the nonzero entry 2 below a_{22} in the second column to be zero.

$$\begin{pmatrix} 1 & -1 & 3 \\ 0 & 2 & -1 \\ 0 & 2 & -5 \end{pmatrix} \xrightarrow{R_2(-1)+R_3} \begin{pmatrix} 1 & -1 & 3 \\ 0 & 2 & -1 \\ 0 & 0 & -4 \end{pmatrix}.$$

The last matrix is a row echelon matrix.

We rewrite the process as follows:

$$A = \begin{pmatrix} 0 & 2 & -1 \\ 1 & -1 & 3 \\ 2 & 0 & 1 \end{pmatrix} \xrightarrow{R_{1,2}} \begin{pmatrix} 1 & -1 & 3 \\ 0 & 2 & -1 \\ 2 & 0 & 1 \end{pmatrix} \xrightarrow{R_1(-2)+R_3}$$
$$\begin{pmatrix} 1 & -1 & 3 \\ 0 & 2 & -1 \\ 0 & 2 & -5 \end{pmatrix} \xrightarrow{R_2(-1)+R_3} \begin{pmatrix} 1 & -1 & 3 \\ 0 & 2 & -1 \\ 0 & 0 & -4 \end{pmatrix}.$$

□

Example 2.12.7. Find a row echelon matrix of

$$A = \begin{pmatrix} 1 & 3 & 0 & 2 & 0 \\ 1 & 1 & 1 & 0 & 0 \\ 2 & 4 & 1 & -1 & 0 \\ 0 & 0 & 0 & 3 & 1 \end{pmatrix}.$$

Solution.

$$A = \begin{pmatrix} 1 & 3 & 0 & 2 & 0 \\ 1 & 1 & 1 & 0 & 0 \\ 2 & 4 & 1 & -1 & 0 \\ 0 & 0 & 0 & 3 & 1 \end{pmatrix} \xrightarrow[R_1(-2)+R_3]{R_1(-1)+R_2} \begin{pmatrix} 1 & 3 & 0 & 2 & 0 \\ 0 & -2 & 1 & -2 & 0 \\ 0 & -2 & 1 & -5 & 0 \\ 0 & 0 & 0 & 3 & 1 \end{pmatrix}$$

$$\xrightarrow{R_2(-1)+R_3} \begin{pmatrix} 1 & 3 & 0 & 2 & 0 \\ 0 & -2 & 1 & -2 & 0 \\ 0 & 0 & 0 & -3 & 0 \\ 0 & 0 & 0 & 3 & 1 \end{pmatrix} \xrightarrow{R_3(1)+R_4}$$

$$\begin{pmatrix} 1 & 3 & 0 & 2 & 0 \\ 0 & -2 & 1 & -2 & 0 \\ 0 & 0 & 0 & -3 & 0 \\ 0 & 0 & 0 & 0 & 1 \end{pmatrix}$$

□

Note that the row echelon matrix of a matrix is not unique. For example, in Example 2.12.7, the row echelon matrix of A is

$$\begin{pmatrix} 1 & 3 & 0 & 2 & 0 \\ 0 & -2 & 1 & -2 & 0 \\ 0 & 0 & 0 & -3 & 0 \\ 0 & 0 & 0 & 0 & 1 \end{pmatrix}.$$

You can continue using row operations to change the row echelon matrix to another row echelon matrix such as

$$\begin{pmatrix} 1 & 3 & 0 & 2 & 0 \\ 0 & -2 & 1 & -2 & 0 \\ 0 & 0 & 0 & -3 & 0 \\ 0 & 0 & 0 & 0 & 1 \end{pmatrix} \xrightarrow{R_3(-\frac{1}{3})} \begin{pmatrix} 1 & 3 & 0 & 2 & 0 \\ 0 & -2 & 1 & -2 & 0 \\ 0 & 0 & 0 & 1 & 0 \\ 0 & 0 & 0 & 0 & 1 \end{pmatrix}.$$

The last matrix is another row echelon matrix of A. However, each matrix has a unique reduced row echelon matrix.

We have the following uniqueness theorem for a reduced row echelon matrix.

Theorem 2.12.8. *Any $m \times n$ matrix can be changed by the row operations to a unique reduced row echelon matrix.*

In the following we shall show the process how to use row operations to find the reduced row echelon matrix of a matrix.

The basic idea is to use the row operations to change a given matrix A to a row echelon matrix B and then use the row operations again to change the row echelon matrix to a reduced row echelon matrix C.

We have studied how to change a matrix to a row echelon matrix, so we first give examples to show how to change a row echelon matrix to a reduced row echelon matrix and then provide examples to show how to change a matrix to a reduced row echelon.

Example 2.12.9. Find the reduced row echelon matrix of

$$B = \begin{pmatrix} 1 & -1 & 3 \\ 0 & 2 & -1 \\ 0 & 0 & -4 \end{pmatrix}.$$

Solution. (1) Change the last leading entry to leading 1 and then use the row operations to change all other entries in the column which contains the leading 1 to be zero.

In the matrix B, the last leading entry is -4, so we use the first row operation to change it to 1 and then change other entries in the column 3 to zero.

$$\begin{aligned} B &= \begin{pmatrix} 1 & -1 & 3 \\ 0 & 2 & -1 \\ 0 & 0 & -4 \end{pmatrix} \xrightarrow{R_3(-\frac{1}{4})} \begin{pmatrix} 1 & -1 & 3 \\ 0 & 2 & -1 \\ 0 & 0 & 1 \end{pmatrix} \\ &\xrightarrow[R_3(-3)+R_1]{R_3(1)+R_2} \begin{pmatrix} 1 & -1 & 0 \\ 0 & 2 & 0 \\ 0 & 0 & 1 \end{pmatrix}. \end{aligned}$$

(2) Use the idea of (1) to other leading entries from lower one to higher ones.

$$\begin{pmatrix} 1 & -1 & 0 \\ 0 & 2 & 0 \\ 0 & 0 & 1 \end{pmatrix} \xrightarrow{R_2(\frac{1}{2})} \begin{pmatrix} 1 & -1 & 0 \\ 0 & 1 & 0 \\ 0 & 0 & 1 \end{pmatrix} \xrightarrow{R_2(1)+R_1} \begin{pmatrix} 1 & 0 & 0 \\ 0 & 1 & 0 \\ 0 & 0 & 1 \end{pmatrix} = I_3.$$

Hence, the reduced row echelon matrix of B is the identity matrix I_3. □

Example 2.12.10. Find the reduced row echelon matrix of

$$B = \begin{pmatrix} 1 & 3 & 0 & 2 & 0 \\ 0 & -2 & 1 & -2 & 0 \\ 0 & 0 & 0 & -3 & 0 \\ 0 & 0 & 0 & 0 & 1 \end{pmatrix}.$$

Solution.

$$B = \begin{pmatrix} 1 & 3 & 0 & 2 & 0 \\ 0 & -2 & 1 & -2 & 0 \\ 0 & 0 & 0 & -3 & 0 \\ 0 & 0 & 0 & 0 & 1 \end{pmatrix} \xrightarrow{R_3(-\frac{1}{3})} \begin{pmatrix} 1 & 3 & 0 & 2 & 0 \\ 0 & -2 & 1 & -2 & 0 \\ 0 & 0 & 0 & 1 & 0 \\ 0 & 0 & 0 & 0 & 1 \end{pmatrix}$$

$$\xrightarrow[R_3(-2)+R_1]{R_3(2)+R_2} \begin{pmatrix} 1 & 3 & 0 & 0 & 0 \\ 0 & -2 & 1 & 0 & 0 \\ 0 & 0 & 0 & 1 & 0 \\ 0 & 0 & 0 & 0 & 1 \end{pmatrix} \xrightarrow{R_2(-\frac{1}{2})}$$

$$\begin{pmatrix} 1 & 3 & 0 & 0 & 0 \\ 0 & 1 & -\frac{1}{2} & 0 & 0 \\ 0 & 0 & 0 & 1 & 0 \\ 0 & 0 & 0 & 0 & 1 \end{pmatrix} \xrightarrow{R_2(-3)+R_1} \begin{pmatrix} 1 & 0 & \frac{3}{2} & 0 & 0 \\ 0 & 1 & -\frac{1}{2} & 0 & 0 \\ 0 & 0 & 0 & 1 & 0 \\ 0 & 0 & 0 & 0 & 1 \end{pmatrix}.$$

The last matrix is the reduced row echelon matrix. □

Combining Examples 2.12.6 and 2.12.9, one can get the reduced row echelon matrix for A given in Example 2.12.6 while combining Examples 2.12.7 and 2.12.10, one can see that the matrix A given in Example 2.12.7 has the reduced row echelon matrix obtained in 2.12.10.

Example 2.12.11. Find the reduced row echelon matrix of

$$A = \begin{pmatrix} 1 & 1 & 2 & 1 \\ 1 & 1 & 1 & -2 \\ 2 & 2 & 1 & 1 \end{pmatrix}.$$

Solution.

$$A = \begin{pmatrix} 1 & 1 & 2 & 1 \\ 1 & 1 & 1 & -2 \\ 2 & 2 & 1 & 1 \end{pmatrix} \xrightarrow[R_1(-2)+R_3]{R_1(-1)+R_2} \begin{pmatrix} 1 & 1 & 2 & 1 \\ 0 & 0 & -1 & -3 \\ 0 & 0 & -3 & -1 \end{pmatrix}$$

$$\xrightarrow{R_2(-3)+R_3} \begin{pmatrix} 1 & 1 & 2 & 1 \\ 0 & 0 & -1 & -3 \\ 0 & 0 & 0 & 8 \end{pmatrix} \xrightarrow{R_3(\frac{1}{8})} \begin{pmatrix} 1 & 1 & 2 & 0 \\ 0 & 0 & -1 & 0 \\ 0 & 0 & 0 & 8 \end{pmatrix}$$

$$\xrightarrow{R_2(-1)} \begin{pmatrix} 1 & 1 & 2 & 0 \\ 0 & 0 & 1 & 0 \\ 0 & 0 & 0 & 1 \end{pmatrix} \xrightarrow{R_2(-2)+R_1} \begin{pmatrix} 1 & 1 & 0 & 0 \\ 0 & 0 & 1 & 0 \\ 0 & 0 & 0 & 1 \end{pmatrix}.$$

The last matrix is the reduced row echelon matrix of A. □

Example 2.12.12. Find the reduced row echelon matrix of

$$A = \begin{pmatrix} 1 & 0 & 1 & 1 & 1 \\ -3 & 1 & 2 & -2 & -5 \\ -2 & -1 & 4 & 8 & 0 \\ 3 & 1 & -3 & -7 & 0 \end{pmatrix}.$$

Solution.

$$A = \begin{pmatrix} 1 & 0 & 1 & 1 & 1 \\ -3 & 1 & 2 & -2 & -5 \\ -2 & -1 & 4 & 8 & 0 \\ 3 & 1 & -3 & -7 & 0 \end{pmatrix} \xrightarrow[R_1(-3)+R_4]{\substack{R_1(3)+R_2 \\ R_1(2)+R_3}} \begin{pmatrix} 1 & 0 & 1 & 1 & 1 \\ 0 & 1 & 5 & 1 & -2 \\ 0 & -1 & 6 & 10 & 2 \\ 0 & 1 & -6 & -10 & -3 \end{pmatrix}$$

$$\xrightarrow[R_2(-1)+R_4]{R_2(1)+R_3} \begin{pmatrix} 1 & 0 & 1 & 1 & 1 \\ 0 & 1 & 5 & 1 & -2 \\ 0 & 0 & 11 & 11 & 0 \\ 0 & 0 & -11 & -11 & -1 \end{pmatrix} \xrightarrow{R_3(1)+R_4}$$

$$\begin{pmatrix} 1 & 0 & 1 & 1 & 1 \\ 0 & 1 & 5 & 1 & -2 \\ 0 & 0 & 11 & 11 & 0 \\ 0 & 0 & 0 & 0 & -1 \end{pmatrix} \xrightarrow[R_3(\frac{1}{11})]{R_4(-1)} \begin{pmatrix} 1 & 0 & 1 & 1 & 1 \\ 0 & 1 & 5 & 1 & -2 \\ 0 & 0 & 1 & 1 & 0 \\ 0 & 0 & 0 & 0 & 1 \end{pmatrix} \xrightarrow[R_4(-1)+R_1]{R_4(2)+R_2}$$

$$\begin{pmatrix} 1 & 0 & 1 & 1 & 0 \\ 0 & 1 & 5 & 1 & 0 \\ 0 & 0 & 1 & 1 & 0 \\ 0 & 0 & 0 & 0 & 1 \end{pmatrix} \xrightarrow[R_3(-1)+R_1]{R_3(-5)+R_2} \begin{pmatrix} 1 & 0 & 0 & 0 & 0 \\ 0 & 1 & 0 & -4 & 0 \\ 0 & 0 & 1 & 1 & 0 \\ 0 & 0 & 0 & 0 & 1 \end{pmatrix}.$$

The last matrix is the reduced row echelon matrix of A. □

Example 2.12.13. Find the reduced row echelon matrix of

$$A = \begin{pmatrix} 2 & 1 & 0 & 1 \\ 3 & 0 & 3 & 1 \\ 4 & 2 & 1 & 3 \end{pmatrix}.$$

Solution.

$$A = \begin{pmatrix} 2 & 1 & 0 & 1 \\ 3 & 0 & 3 & 1 \\ 4 & 2 & 1 & 3 \end{pmatrix} \xrightarrow[R_1(-2)+R_3]{R_1(-1)+R_2} \begin{pmatrix} 2 & 1 & 0 & 1 \\ 1 & -1 & 3 & 0 \\ 0 & 0 & 1 & 1 \end{pmatrix} \xrightarrow{R_{1,2}}$$

$$\begin{pmatrix} 1 & -1 & 3 & 0 \\ 2 & 1 & 0 & 1 \\ 0 & 0 & 1 & 1 \end{pmatrix} \xrightarrow{R_1(-2)+R_2} \begin{pmatrix} 1 & -1 & 3 & 0 \\ 0 & 3 & -6 & 1 \\ 0 & 0 & 1 & 1 \end{pmatrix} \xrightarrow[R_3(6)+R_2]{R_3(-3)+R_1}$$

$$\begin{pmatrix} 1 & -1 & 0 & -3 \\ 0 & 3 & 0 & 7 \\ 0 & 0 & 1 & 1 \end{pmatrix} \xrightarrow{R_2(\frac{1}{3})} \begin{pmatrix} 1 & -1 & 0 & -3 \\ 0 & 1 & 0 & \frac{7}{3} \\ 0 & 0 & 1 & 1 \end{pmatrix} \xrightarrow{R_2(1)+R_1}$$

$$\begin{pmatrix} 1 & 0 & 0 & -\frac{2}{3} \\ 0 & 1 & 0 & \frac{7}{3} \\ 0 & 0 & 1 & 1 \end{pmatrix}.$$

The last matrix is the reduced row echelon matrix of A. □

We use the row operation $R_1(-1)+R_2$ to change row 2 in order to make the first number in the second row to be 1 and then interchange the new second row and the first row to get the first number in the first row to be 1. If you use $R_1(\frac{1}{2})$ to A first, then you will meet many calculations of fractions.

Exercises

1. Let $A = \begin{pmatrix} 1 & 0 & 1 \\ 2 & 1 & -1 \\ -3 & 2 & -2 \end{pmatrix}$. Use the second row operation to make the numbers 2 and -3 in A to be zero.

2. Let $A = \begin{pmatrix} 1 & 0 & 1 \\ 0 & 1 & -3 \\ 0 & 3 & 1 \end{pmatrix}$. Use the second row operation to make the number 3 in A to be zero.

3. Find a row echelon matrix for each of the following matrices

$$A = \begin{pmatrix} 1 & 0 & 1 \\ 2 & 1 & -1 \\ -3 & 2 & -2 \end{pmatrix} \quad B = \begin{pmatrix} 0 & 1 & -2 \\ 1 & -1 & 1 \\ -2 & 1 & 0 \end{pmatrix}$$

4. Find the reduced row echelon matrix of each of the following matrices.

$$A = \begin{pmatrix} 2 & -4 & 2 \\ 0 & 1 & -3 \\ 0 & 0 & -6 \end{pmatrix} \quad B = \begin{pmatrix} 1 & 1 & 0 & 1 & 0 \\ 0 & 3 & 6 & -3 & -6 \\ 0 & 0 & 0 & -2 & 2 \\ 0 & 0 & 0 & 0 & 2 \end{pmatrix}$$

$$C = \begin{pmatrix} 1 & 0 & 0 & -1 \\ 1 & -2 & 1 & 1 \\ 2 & -2 & 1 & 2 \end{pmatrix} \qquad D = \begin{pmatrix} 0 & 0 & 1 & 1 & -1 \\ -1 & 1 & 2 & -2 & -1 \\ -2 & -2 & 4 & 0 & 0 \\ 2 & -1 & 0 & 1 & 0 \end{pmatrix}$$

$$E = \begin{pmatrix} 2 & 12 & 0 & -1 \\ 3 & 0 & 3 & 2 \\ -3 & 2 & -1 & 6 \end{pmatrix}$$

2.13 Ranks and nullity of matrices

Definition 2.13.1. Let A be an $m \times n$ matrix and let B be one of its row echelon matrices. The number of nonzero rows in the row echelon matrix B of A is called the rank of A, denoted by $r(A)$. $n - r(A)$ is called the nullity of A, denoted by $\text{null}(A)$, that is,

$$\text{null}(A) = n - r(A). \tag{2.17}$$

According to Definition 2.13.1, the rank of A is equal to the rank of any of its row echelon matrices.

The method to find the rank of a matrix is to use the row operations to change A to a row echelon matrix and then count the numbers of nonzero rows which is the rank of A.

Example 2.13.2. Find the ranks and nullities of the following matrices.

$$A = \begin{pmatrix} 1 & 0 & 0 & 0 \\ 0 & 1 & 0 & 0 \\ 0 & 0 & 0 & 0 \end{pmatrix} \quad B = \begin{pmatrix} 1 & 2 & -1 \\ 2 & -1 & 3 \end{pmatrix} \quad C = \begin{pmatrix} 0 & 4 & 2 & 1 \\ 0 & 2 & 1 & 1 \\ 1 & 1 & 1 & 4 \\ 2 & 0 & 1 & 3 \end{pmatrix}$$

Solution. Since A itself is a reduced row echelon matrix, so we do not need to use any row operations. Note that A has two nonzero rows, so the rank of A is 2, that is $r(A) = 2$. Since A is a 3×4 matrix, so $\text{null}(A) = 4 - r(A) = 2$.

Since B is not a row echelon matrix, we need to use the row operations to find its row echelon matrix.

$$B = \begin{pmatrix} 1 & 2 & -1 \\ 2 & -1 & 3 \end{pmatrix} \xrightarrow{R_1(-2) + R_2} \begin{pmatrix} 1 & 2 & -1 \\ 0 & -5 & 5 \end{pmatrix}.$$

The last matrix is a row echelon matrix of B and there are two nonzero rows. Hence, $r(B) = 2$. Since B is a 2×3 matrix, so $\text{null}(B) = 3 - 2 = 1$.

We need to use row operations to change C to its row echelon matrix.

$$C = \begin{pmatrix} 0 & 4 & 2 & 1 \\ 0 & 2 & 1 & 1 \\ 1 & 1 & 1 & 4 \\ 2 & 0 & 1 & 3 \end{pmatrix} \xrightarrow{R_{1,3}} \begin{pmatrix} 1 & 1 & 1 & 4 \\ 0 & 2 & 1 & 1 \\ 0 & 4 & 2 & 1 \\ 2 & 0 & 1 & 3 \end{pmatrix} \xrightarrow{R_1(-2)+R_4}$$

$$\begin{pmatrix} 1 & 1 & 1 & 4 \\ 0 & 2 & 1 & 1 \\ 0 & 4 & 2 & 1 \\ 0 & -2 & -1 & -5 \end{pmatrix} \xrightarrow[R_2(1)+R_4]{R_2(-2)+R_3} \begin{pmatrix} 1 & 1 & 1 & 4 \\ 0 & 2 & 1 & 1 \\ 0 & 0 & 0 & -1 \\ 0 & 0 & 0 & -4 \end{pmatrix}$$

$$\xrightarrow{R_3(-4)+R_4} \begin{pmatrix} 1 & 1 & 1 & 4 \\ 0 & 2 & 1 & 1 \\ 0 & 0 & 0 & -1 \\ 0 & 0 & 0 & 0 \end{pmatrix}.$$

The last matrix is a row echelon matrix of C and there are three nonzero rows. Hence, $r(C) = 3$ and $\text{null}(C) = 4 - 3 = 1$. □

The row echelon matrix of a matrix A is not unique, but all row echelon matrices of A have same numbers of nonzero rows. Hence, no matter which row operations are used to change A to a row echelon matrix, the rank of A keeps unchanged, that is, the rank $r(A)$ is unique.

Now, we consider the rank and nullity of A^T.

The following result shows that the rank of A^T is equal to the rank of A.

Theorem 2.13.3. *Let A be an $m \times n$ matrix. Then*

(1) $r(A^T) = r(A)$;

(2) $null(A^T) = m - r(A)$.

For some matrices, you may find the rank of A^T first and then use the Theorem 2.13.3 to get the rank of A.

Example 2.13.4. Find the rank of $A = \begin{pmatrix} 1 & 0 & 0 & 0 \\ 1 & 2 & 0 & 0 \\ 1 & 1 & 1 & -1 \\ 2 & -1 & 0 & 0 \end{pmatrix}$.

Solution.

$$A^T = \begin{pmatrix} 1 & 1 & 1 & 2 \\ 0 & 2 & 1 & -1 \\ 0 & 0 & 1 & 0 \\ 0 & 0 & -1 & 0 \end{pmatrix} \xrightarrow{R_3(1)+R_4} \begin{pmatrix} 1 & 1 & 1 & 2 \\ 0 & 2 & 1 & -1 \\ 0 & 0 & 1 & 0 \\ 0 & 0 & 0 & 0 \end{pmatrix}.$$

The last matrix is a row echelon matrix and contains 3 nonzero rows. Hence, $r(A^T) = 3$. By Theorem 2.13.3, $r(A) = r(A^T) = 3$. □

In Example 2.13.4, if we use row operations to A to find the rank of A, we need to use row operations 5 times while we only use one row operation to A^T to get the rank of A.

The following result gives the relation between the rank of a matrix and its size.

Theorem 2.13.5. *Let A be an $m \times n$ matrix. Then*

$$r(A) \leq \min\{m, n\}.$$

Exercises

1. Find the ranks and nullities of the following matrices.

$$A = \begin{pmatrix} 1 & 0 & 0 & 1 \\ 0 & -1 & 0 & 0 \\ 0 & 0 & 0 & 0 \end{pmatrix} \quad B = \begin{pmatrix} 0 & 2 & -1 \\ 2 & -1 & 3 \end{pmatrix} \quad C = \begin{pmatrix} 1 & 4 & 2 & 1 \\ 0 & 2 & 1 & 1 \\ 1 & 0 & 1 & 4 \\ 2 & 0 & 1 & -3 \end{pmatrix}$$

2.14 The inverse of a matrix

Definition 2.14.1. Let A and B be square matrices of order n. Suppose

$$AB = I_n.$$

Then A is said to be invertible and B is called the inverse of A.

We write $B = A^{-1}$. If A is not invertible, then A is said to be singular. By Definition 2.14.1, if A is invertible, then

$$AA^{-1} = I_n.$$

A is not invertible if and only if for each square matrix B of order n,

$$AB \neq I_n$$

The following theorem provides some equivalent results of inverse matrices.

Theorem 2.14.2. *Let A and B are square matrices of order n. Then the following are equivalent.*

(i) $AB = I_n$.
(ii) $BA = I_n$.
(iii) $AB = BA = I_n$.

Example 2.14.3. Show that B is an inverse of A if

$$A = \begin{pmatrix} 2 & -5 \\ -1 & 3 \end{pmatrix} \text{ and } B = \begin{pmatrix} 3 & 5 \\ 1 & 2 \end{pmatrix}.$$

Solution.

$$AB = \begin{pmatrix} 2 & -5 \\ -1 & 3 \end{pmatrix}\begin{pmatrix} 3 & 5 \\ 1 & 2 \end{pmatrix} = \begin{pmatrix} 1 & 0 \\ 0 & 1 \end{pmatrix} = I_2.$$

Hence, B is an inverse of A. □

Example 2.14.4. Let $A = \begin{pmatrix} a_1 & 0 & \cdots & 0 \\ 0 & a_2 & \cdots & 0 \\ \vdots & \vdots & \ddots & \vdots \\ 0 & 0 & \cdots & a_n \end{pmatrix}$. Assume that $a_i \neq 0$ for each $i \in \mathcal{I}_n$. Show that

$$A^{-1} = \begin{pmatrix} \frac{1}{a_1} & 0 & \cdots & 0 \\ 0 & \frac{1}{a_2} & \cdots & 0 \\ \vdots & \vdots & \ddots & \vdots \\ 0 & 0 & \cdots & \frac{1}{a_n} \end{pmatrix}.$$

Solution. Since

$$\begin{aligned} AA^{-1} &= \begin{pmatrix} a_1 & 0 & \cdots & 0 \\ 0 & a_2 & \cdots & 0 \\ \vdots & \vdots & \ddots & \vdots \\ 0 & 0 & \cdots & a_n \end{pmatrix}\begin{pmatrix} \frac{1}{a_1} & 0 & \cdots & 0 \\ 0 & \frac{1}{a_2} & \cdots & 0 \\ \vdots & \vdots & \ddots & \vdots \\ 0 & 0 & \cdots & \frac{1}{a_n} \end{pmatrix} \\ &= \begin{pmatrix} 1 & 0 & \cdots & 0 \\ 0 & 1 & \cdots & 0 \\ \vdots & \vdots & \ddots & \vdots \\ 0 & 0 & \cdots & 0 \end{pmatrix}, \end{aligned}$$

the inverse of A is A^{-1}. □

Example 2.14.5. Let $A = \begin{pmatrix} \frac{1}{2} & 0 & 0 \\ 0 & \frac{1}{3} & 0 \\ 0 & 0 & 4 \end{pmatrix}$. Find A^{-1}.

Solution. By Example 2.14.4, we have $A^{-1} = \begin{pmatrix} 2 & 0 & 0 \\ 0 & 3 & 0 \\ 0 & 0 & \frac{1}{4}. \end{pmatrix}$. □

Now, we show how to find the inverse matrices of 2×2 matrices.
Let

$$A = \begin{pmatrix} a_{11} & a_{12} \\ a_{21} & a_{22} \end{pmatrix}. \tag{2.18}$$

We define the determinant of A as follows:

$$|A| = a_{11}a_{22} - a_{12}a_{21}. \tag{2.19}$$

Theorem 2.14.6. *Let A be same as in* (2.18). *Then the following assertions hold.*

(1) *A is invertible if and only if $|A| \neq 0$.*
(2) *If A is invertible, then*

$$A^{-1} = \frac{1}{|A|}\begin{pmatrix} a_{22} & -a_{12} \\ -a_{21} & a_{11} \end{pmatrix}. \tag{2.20}$$

Example 2.14.7. Let $A = \begin{pmatrix} 2 & -4 \\ 1 & 3 \end{pmatrix}$. Determine if A is invertible. If so, calculate A^{-1}.

Solution. Since

$$|A| = \begin{vmatrix} 2 & -4 \\ 1 & 3 \end{vmatrix} = (2)(3) - (-4)(1) = 10 \neq 0.$$

By Theorem 2.14.6, A is invertible and

$$A^{-1} = \frac{1}{|A|}\begin{pmatrix} 3 & 4 \\ -1 & 2 \end{pmatrix} = \frac{1}{10}\begin{pmatrix} 3 & 4 \\ -1 & 2 \end{pmatrix} = \begin{pmatrix} \frac{3}{10} & \frac{2}{5} \\ -\frac{1}{10} & \frac{1}{5} \end{pmatrix}.$$

□

Example 2.14.8. Let $A = \begin{pmatrix} 1 & 2 \\ -2 & -4 \end{pmatrix}$. Determine if A is invertible. If so, calculate A^{-1}.

Solution. Since

$$|A| = \begin{vmatrix} 1 & 2 \\ -2 & -4 \end{vmatrix} = (1)(-4) - (2)(-2) = 0,$$

A is not invertible and A^{-1} does not exist. □

Example 2.14.9. Let $A = \begin{pmatrix} 1 & x^3 \\ 1 & 1 \end{pmatrix}$. Find all $x \in \mathbb{R}$ such that A is invertible.

Solution. By computation, we have

$$|A| = \begin{vmatrix} 1 & x^3 \\ 1 & 1 \end{vmatrix} = 1 - (x^3)(1) = 1 - x^3.$$

By Theorem 2.14.6, if $|A| \neq 0$, then A is invertible. Hence, if $1 - x^3 \neq 0$, that is, $x \neq 1$, then A is invertible. □

Example 2.14.10. Let $A = \begin{pmatrix} x & 4 \\ 1 & x \end{pmatrix}$. Find all $x \in \mathbb{R}$ such that A is not invertible.

Solution. By computation, we have

$$|A| = \begin{vmatrix} x & 4 \\ 1 & x \end{vmatrix} = (x)(x) - (4)(1) = x^2 - 4.$$

By Theorem 2.14.6, if $|A| = 0$, then A is not invertible. Hence, if $x^2 - 4 = 0$, that is, $x = 2$ or $x = -2$, then A is not invertible. □

Example 2.14.11. Let $A = \begin{pmatrix} 0 & 0 \\ 0 & 0 \end{pmatrix}$. Show that A is not invertible.

Solution. Since $|A| = \begin{vmatrix} 0 & 0 \\ 0 & 0 \end{vmatrix} = 0$, A is not invertible. □

The method of finding inverse matrices given in Theorem 2.14.6 is applied only to 2×2 matrices. For square matrices of order $n > 2$, we have to use another method involving row operations to find their inverses.

Let A be a square matrix of order $n \geq 2$. Consider the following matrix

$$(A|I_n). \tag{2.21}$$

(*i*) If you can use row operations to change $(A|I_n)$ to a matrix $(I_n|B)$, then $A^{-1} = B$.

(*ii*) If you can use row operations to change $(A|I_n)$ to a matrix $(C|D)$, where C contains at lease one zero row, then A is not invertible.

Example 2.14.12. Let $A = \begin{pmatrix} 2 & -3 \\ -4 & 5 \end{pmatrix}$. Determine if A is invertible. If so, calculate A^{-1}.

Solution.

$$(A|I_2) = \left(\begin{array}{cc|cc} 2 & -3 & 1 & 0 \\ -4 & 5 & 0 & 1 \end{array}\right) \xrightarrow{R_1(\frac{1}{2})} \left(\begin{array}{cc|cc} 1 & -\frac{3}{2} & \frac{1}{2} & 0 \\ -4 & 5 & 0 & 1 \end{array}\right) \xrightarrow{R_1(4)+R_2}$$

$$\left(\begin{array}{cc|cc} 1 & -\frac{3}{2} & \frac{1}{2} & 0 \\ 0 & -1 & 2 & 1 \end{array}\right) \xrightarrow{R_2(-1)} \left(\begin{array}{cc|cc} 1 & -\frac{3}{2} & \frac{1}{2} & 0 \\ 0 & 1 & -2 & -1 \end{array}\right)$$

$$\xrightarrow{R_2(\frac{3}{2})+R_1} \left(\begin{array}{cc|cc} 1 & 0 & -\frac{5}{2} & -\frac{3}{2} \\ 0 & 1 & -2 & -1 \end{array}\right) = (I_2|B).$$

Hence, A is invertible and $A^{-1} = \begin{pmatrix} -\frac{5}{2} & -\frac{3}{2} \\ -2 & -1 \end{pmatrix}$. □

Example 2.14.13. Let $A = \begin{pmatrix} 1 & 1 & 1 \\ 0 & 1 & 1 \\ 1 & 0 & 1 \end{pmatrix}$. Find the inverse of A.

Solution.

$$(A|I_3) = \left(\begin{array}{ccc|ccc} 1 & 1 & 1 & 1 & 0 & 0 \\ 0 & 1 & 1 & 0 & 1 & 0 \\ 1 & 0 & 1 & 0 & 0 & 1 \end{array}\right) \xrightarrow{R_1(-1)+R_3}$$

$$\left(\begin{array}{ccc|ccc} 1 & 1 & 1 & 1 & 0 & 0 \\ 0 & 1 & 1 & 0 & 1 & 0 \\ 0 & -1 & 0 & -1 & 0 & 1 \end{array}\right) \xrightarrow{R_2(1)+R_3}$$

$$\left(\begin{array}{ccc|ccc} 1 & 1 & 1 & 1 & 0 & 0 \\ 0 & 1 & 1 & 0 & 1 & 0 \\ 0 & 0 & 1 & -1 & 1 & 1 \end{array}\right) \xrightarrow[R_3(-1)+R_2]{R_3(-1)+R_1}$$

$$\left(\begin{array}{ccc|ccc} 1 & 1 & 0 & 2 & -1 & -1 \\ 0 & 1 & 0 & 1 & 0 & -1 \\ 0 & 0 & 1 & -1 & 1 & 1 \end{array}\right) \xrightarrow{R_2(-1)+R_1}$$

$$\left(\begin{array}{ccc|ccc} 1 & 0 & 0 & 1 & -1 & 0 \\ 0 & 1 & 0 & 1 & 0 & -1 \\ 0 & 0 & 1 & -1 & 1 & 1 \end{array}\right) = (I_3|A^{-1}).$$

Hence A is invertible and $A^{-1} = \begin{pmatrix} 1 & -1 & 0 \\ 1 & 0 & -1 \\ -1 & 1 & 1 \end{pmatrix}$. □

Note that you can verify whether the inverse matrix you find is correct by checking if $AA^{-1} = I_3$.

Example 2.14.14. Let $A = \begin{pmatrix} 1 & -3 & 4 \\ 2 & -5 & 7 \\ 0 & -1 & 1 \end{pmatrix}$. Calculate A^{-1} if it exists.

Solution.

$$
\begin{aligned}
(A|I_3) &= \left(\begin{array}{ccc|ccc} 1 & -3 & 4 & 1 & 0 & 0 \\ 2 & -5 & 7 & 0 & 1 & 0 \\ 0 & -1 & 1 & 0 & 0 & 1 \end{array}\right) \xrightarrow{R_1(-2)+R_2} \\
&\left(\begin{array}{ccc|ccc} 1 & -3 & 4 & 1 & 0 & 0 \\ 0 & 1 & -1 & -2 & 1 & 0 \\ 0 & -1 & 1 & 0 & 0 & 1 \end{array}\right) \xrightarrow[R_2(1)+R_3]{R_2(3)+R_1} \\
&\left(\begin{array}{ccc|ccc} 1 & 0 & 1 & -5 & 3 & 0 \\ 0 & 1 & -1 & -2 & 1 & 0 \\ 0 & 0 & 0 & -2 & 1 & 1 \end{array}\right) = (C|D).
\end{aligned}
$$

Since $C = \begin{pmatrix} 1 & 0 & 1 \\ 0 & 1 & -1 \\ 0 & 0 & 0 \end{pmatrix}$ contains a row of zero, that is, the third row of C is zero. Hence, A is not invertible and A^{-1} does not exist. □

Example 2.14.15. Let $A = \begin{pmatrix} 1 & 0 & 0 & 0 \\ 0 & 2 & 0 & 0 \\ 0 & 0 & 3 & 0 \\ 0 & 0 & 0 & 4 \end{pmatrix}$. Find A^{-1} if it exists.

Solution. Since

$$
\begin{aligned}
(A|I_4) &= \left(\begin{array}{cccc|cccc} 1 & 0 & 0 & 0 & 1 & 0 & 0 & 0 \\ 0 & 2 & 0 & 0 & 0 & 1 & 0 & 0 \\ 0 & 0 & 3 & 0 & 0 & 0 & 1 & 0 \\ 0 & 0 & 0 & 4 & 0 & 0 & 0 & 1 \end{array}\right) \xrightarrow[R_4(\frac{1}{4})]{\substack{R_2(\frac{1}{2}) \\ R_3(\frac{1}{3})}} \\
&\left(\begin{array}{cccc|cccc} 1 & 0 & 0 & 0 & 1 & 0 & 0 & 0 \\ 0 & 1 & 0 & 0 & 0 & \frac{1}{2} & 0 & 0 \\ 0 & 0 & 1 & 0 & 0 & 0 & \frac{1}{3} & 0 \\ 0 & 0 & 0 & 1 & 0 & 0 & 0 & \frac{1}{4} \end{array}\right) = (I_4|A^{-1}),
\end{aligned}
$$

Hence, we obtain $A^{-1} = \begin{pmatrix} 1 & 0 & 0 & 0 \\ 0 & \frac{1}{2} & 0 & 0 \\ 0 & 0 & \frac{1}{3} & 0 \\ 0 & 0 & 0 & \frac{1}{4} \end{pmatrix}$. □

The following theorem provides some properties of inverse matrices.

Theorem 2.14.16. *Let A and B be $n \times n$ matrices. Then the following assertions hold.*

(1) *If A is invertible, then $(A^{-1})^{-1} = A$.*

(2) *If A is invertible, then its inverse is unique.*

(3) *If A and B are invertible, then AB is invertible and*

$$(AB)^{-1} = B^{-1}A^{-1}.$$

(4) *If A is invertible and $k \neq 0$, then kA is invertible and*

$$(kA)^{-1} = k^{-1}A^{-1}.$$

(5) *If A is invertible, then A^n is invertible for $n \in \mathbb{N}$ and*

$$(A^n)^{-1} = (A^{-1})^n.$$

In this case, we write $A^{-n} = (A^{-1})^n$

(6) *If A is invertible, then A^T is invertible and*

$$(A^T)^{-1} = (A^{-1})^T.$$

(7) *If A is invertible and symmetric, then A^{-1} is symmetric.*

(8) *If A is invertible, then AA^T and A^TA are also invertible.*

(9) *Assume that A is triangular, then A is invertible if and only if its diagonal entries are all nonzero.*

(10) *If A is lower triangular and invertible, then A^{-1} is lower triangular and invertible. If A is upper triangular and invertible, then A^{-1} is upper triangular and invertible.*

Example 2.14.17. Let $A = \begin{pmatrix} 1 & 2 \\ 1 & 3 \end{pmatrix}$ and $B = \begin{pmatrix} 3 & 2 \\ 2 & 2 \end{pmatrix}$.

(1) Find A^{-1}, B^{-1} and $(AB)^{-1}$.
(2) Verify $(AB)^{-1} = B^{-1}A^{-1}$.

Solution. Since $|A| = \begin{vmatrix} 1 & 2 \\ 1 & 3 \end{vmatrix} = 3 - 2 = 1$,

$$A^{-1} = \frac{1}{|A|}\begin{pmatrix} d & -b \\ -c & a \end{pmatrix} = \begin{pmatrix} 3 & -2 \\ -1 & 1 \end{pmatrix}.$$

Similarly, since $|B| = \begin{vmatrix} 3 & 2 \\ 2 & 2 \end{vmatrix} = 6 - 4 = 2$,

$$B^{-1} = \frac{1}{|B|}\begin{pmatrix} d & -b \\ -c & a \end{pmatrix} = \frac{1}{2}\begin{pmatrix} 2 & -2 \\ -2 & 3 \end{pmatrix} = \begin{pmatrix} 1 & -1 \\ -1 & \frac{3}{2} \end{pmatrix}.$$

Moreover, $B^{-1}A^{-1} = \begin{pmatrix} 1 & -1 \\ -1 & \frac{3}{2} \end{pmatrix} \begin{pmatrix} 3 & -2 \\ -1 & 1 \end{pmatrix} = \begin{pmatrix} 4 & -3 \\ -\frac{9}{2} & \frac{7}{2} \end{pmatrix}$.

Since

$$AB = \begin{pmatrix} 1 & 2 \\ 1 & 3 \end{pmatrix} \begin{pmatrix} 3 & 2 \\ 2 & 2 \end{pmatrix} = \begin{pmatrix} 7 & 6 \\ 9 & 8 \end{pmatrix}$$

and $|AB| = \begin{vmatrix} 7 & 6 \\ 9 & 8 \end{vmatrix} = 56 - 54 = 2$, we have

$$(AB)^{-1} = \frac{1}{2} \begin{pmatrix} 8 & -6 \\ -9 & 7 \end{pmatrix} = \begin{pmatrix} 4 & -3 \\ -\dfrac{9}{2} & \dfrac{7}{2} \end{pmatrix}.$$

Hence, we obtain $(AB)^{-1} = B^{-1}A^{-1}$. □

Example 2.14.18. Let $A = \begin{pmatrix} 1 & 2 \\ 1 & 3 \end{pmatrix}$. Find A^3, A^{-1}, $(A^{-1})^3$ and verify $(A^3)^{-1} = (A^{-1})^3$.

Solution. $A^3 = \begin{pmatrix} 1 & 2 \\ 1 & 3 \end{pmatrix} \begin{pmatrix} 1 & 2 \\ 1 & 3 \end{pmatrix} \begin{pmatrix} 1 & 2 \\ 1 & 3 \end{pmatrix} = \begin{pmatrix} 11 & 30 \\ 15 & 41 \end{pmatrix}$. It is easy to verify that $A^{-1} = \begin{pmatrix} 3 & -2 \\ -1 & 1 \end{pmatrix}$. Hence,

$$(A^{-1})^3 = \begin{pmatrix} 3 & -2 \\ -1 & 1 \end{pmatrix} \begin{pmatrix} 3 & -2 \\ -1 & 1 \end{pmatrix} \begin{pmatrix} 3 & -2 \\ -1 & 1 \end{pmatrix} = \begin{pmatrix} 41 & -30 \\ -15 & 11 \end{pmatrix}.$$

Since $|A^3| = (11)(41) - (30)(15) = 1$,

$$(A^3)^{-1} = \frac{1}{|A^3|} \begin{pmatrix} 41 & -30 \\ -15 & 11 \end{pmatrix} = \begin{pmatrix} 41 & -30 \\ -15 & 11 \end{pmatrix}.$$

Hence, $(A^3)^{-1} = (A^{-1})^3$. □

Example 2.14.19. Let $A = \begin{pmatrix} -5 & -3 \\ 2 & 1 \end{pmatrix}$. Find A^{-1}, $(A^T)^{-1}$ and verify that $(A^T)^{-1} = (A^{-1})^T$.

Solution. Since $|A| = (-5)(1) - (-3)(2) = 1$,

$$A^{-1} = \frac{1}{|A|} \begin{pmatrix} 1 & 3 \\ -2 & -5 \end{pmatrix} = \begin{pmatrix} 1 & 3 \\ -2 & -5 \end{pmatrix}.$$

and $(A^{-1})^T = \begin{pmatrix} 1 & -2 \\ 3 & -5 \end{pmatrix}$. Since $A^T = \begin{pmatrix} -5 & 2 \\ -3 & 1 \end{pmatrix}$, $(A^T)^{-1} = \begin{pmatrix} 1 & -2 \\ 3 & -5 \end{pmatrix}$.
Hence, $(A^T)^{-1} = (A^{-1})^T$. □

Example 2.14.20. Let $A = \begin{pmatrix} 1 & 3 \\ 3 & 2 \end{pmatrix}$. Find A^{-1}. Is A^{-1} symmetric.

Solution. Since $|A| = 2 - 9 = -7$,

$$A^{-1} = \frac{1}{|A|}\begin{pmatrix} 2 & -3 \\ -3 & 1 \end{pmatrix} = -\frac{1}{7}\begin{pmatrix} 2 & -3 \\ -3 & 1 \end{pmatrix} = \begin{pmatrix} -\frac{2}{7} & \frac{3}{7} \\ \frac{3}{7} & -\frac{1}{7} \end{pmatrix}$$

and A^{-1} is symmetric. □

Example 2.14.21. Determine whether the following triangular matrices are invertible.

$$A = \begin{pmatrix} 1 & 0 & 0 \\ 2 & 3 & 0 \\ 3 & 0 & 1 \end{pmatrix} \quad B = \begin{pmatrix} 2 & 0 & 0 \\ 1 & 0 & 0 \\ 0 & 1 & 3 \end{pmatrix} \quad C = \begin{pmatrix} 4 & 0 & 0 \\ 0 & 2 & 1 \\ 0 & 0 & 5 \end{pmatrix} \quad D = \begin{pmatrix} 1 & 2 & 0 \\ 0 & 0 & 3 \\ 0 & 0 & 6 \end{pmatrix}$$

Solution. A and C are invertible since $|A| = 3 \neq 0$ and $|C| = 40 \neq 0$. B and D are singular since $|B| = 0$ and $|D| = 0$. □

Exercises

1. Let $A = \begin{pmatrix} \frac{1}{3} & 0 & 0 \\ 0 & \frac{1}{4} & 0 \\ 0 & 0 & 4 \end{pmatrix}$. Find A^{-1}.

2. Let $A = \begin{pmatrix} 3 & -4 \\ 2 & 3 \end{pmatrix}$. Determine if A is invertible. If so, calculate A^{-1}.

3. Let $A = \begin{pmatrix} 2 & 4 \\ -1 & -2 \end{pmatrix}$. Determine if A is invertible. If so, calculate A^{-1}.

4. Let $A = \begin{pmatrix} 1 & x^2 \\ 1 & 1 \end{pmatrix}$. Find all $x \in \mathbb{R}$ such that A is invertible.

5. Let $A = \begin{pmatrix} x & 3 \\ 3 & x \end{pmatrix}$. Find all $x \in \mathbb{R}$ such that A is not invertible.

6. Let $A = \begin{pmatrix} 1 & -3 \\ -4 & 5 \end{pmatrix}$. Use the row operations to determine if A is invertible. If so, calculate A^{-1}.

7. Use the row operations to determine if the following matrix is invertible. If so, calculate its inverse.

$$A = \begin{pmatrix} 1 & -1 & 1 \\ 0 & 1 & -1 \\ 2 & 0 & -1 \end{pmatrix} \qquad B = \begin{pmatrix} 1 & 0 & 1 & 1 \\ 0 & 2 & 1 & 6 \\ -4 & 0 & -3 & -3 \\ 2 & 1 & -3 & 0 \end{pmatrix}$$

8. Let $A = \begin{pmatrix} 1 & 0 \\ 2 & 3 \end{pmatrix}$ and $B = \begin{pmatrix} -1 & 2 \\ 2 & 1 \end{pmatrix}$. Find A^{-1}, B^{-1}, $(AB)^{-1}$ and verify $(AB)^{-1} = B^{-1}A^{-1}$.

9. Let $A = \begin{pmatrix} -1 & 2 \\ 1 & 3 \end{pmatrix}$. Find A^3, A^{-1}, $(A^{-1})^3$ and verify

$$(A^3)^{-1} = (A^{-1})^3.$$

10. Let $A = \begin{pmatrix} -4 & 1 \\ 3 & 1 \end{pmatrix}$. Find A^{-1}, $(A^T)^{-1}$ and verify that

$$(A^T)^{-1} = (A^{-1})^T.$$

11. Let $A = \begin{pmatrix} 2 & 2 \\ 2 & 1 \end{pmatrix}$. Find A^{-1}. Is A^{-1} symmetric.

12. Determine whether the following triangular matrices are invertible.

$$A = \begin{pmatrix} 1 & 0 & 0 \\ 2 & 1 & 0 \\ 3 & 0 & -1 \end{pmatrix} \qquad B = \begin{pmatrix} 0 & 0 & 0 \\ 1 & 1 & 0 \\ 0 & 1 & 2 \end{pmatrix} \qquad C = \begin{pmatrix} 1 & 0 & 0 \\ 0 & 3 & 1 \\ 0 & 0 & 4 \end{pmatrix}$$

$$D = \begin{pmatrix} 5 & 3 & 0 \\ 0 & 1 & 3 \\ 0 & 0 & 0 \end{pmatrix}$$

Chapter 3

Determinants

3.1 Determinants of 2×2 matrices

Let

$$A = \begin{pmatrix} a_{11} & a_{12} \\ a_{21} & a_{22} \end{pmatrix}.$$

Recall that the determinant of A denoted by $|A|$, is defined by the number

$$|A| = \begin{vmatrix} a_{11} & a_{12} \\ a_{21} & a_{22} \end{vmatrix} = a_{11}a_{22} - a_{12}a_{21}.$$

We remark that the two numbers a_{11}, a_{12} in the first term and the two numbers a_{21}, a_{22} in the second term are from different rows and different columns.

Example 3.1.1. Evaluate determinants of the following matrices.

$$A = \begin{pmatrix} 3 & 1 \\ 4 & -2 \end{pmatrix} \quad B = \begin{pmatrix} 2 & 0 \\ 1 & 3 \end{pmatrix} \quad C = \begin{pmatrix} 4 & 1 \\ 0 & 3 \end{pmatrix} \quad D = \begin{pmatrix} 2 & 3 \\ 4 & 6 \end{pmatrix}$$

Solution.

$$|A| = \begin{vmatrix} 3 & 1 \\ 4 & -2 \end{vmatrix} = (3)(-2) - (1)(4) = -6 - 4 = -10.$$

$$|B| = \begin{vmatrix} 2 & 0 \\ 1 & 3 \end{vmatrix} = (2)(3) - (0)(1) = 6.$$

$$|C| = \begin{vmatrix} 4 & 1 \\ 0 & 3 \end{vmatrix} = (4)(3) - (1)(0) = 12.$$

$$|D| = \begin{vmatrix} 2 & 3 \\ 4 & 6 \end{vmatrix} = (2)(6) - (3)(4) = 0.$$

□

From the above example, we see that the determinants can be negative, zero or positive.

Exercises

1. Evaluate each of the following determinants.

$$A = \begin{pmatrix} 1 & 1 \\ 4 & -2 \end{pmatrix} B = \begin{pmatrix} -2 & 2 \\ 1 & -3 \end{pmatrix} C = \begin{pmatrix} 2 & -1 \\ 3 & 0 \end{pmatrix} D = \begin{pmatrix} -2 & 1 \\ -6 & 3 \end{pmatrix}$$

3.2 Determinants of 3×3 matrices

Let

$$A = \begin{pmatrix} a_{11} & a_{12} & a_{13} \\ a_{21} & a_{22} & a_{23} \\ a_{31} & a_{32} & a_{33} \end{pmatrix}.$$

The determinant of A, denoted by $|A|$, is defined by the number

$$\begin{aligned} |A| &= \begin{vmatrix} a_{11} & a_{12} & a_{13} \\ a_{21} & a_{22} & a_{23} \\ a_{31} & a_{32} & a_{33} \end{vmatrix} = (a_{11}a_{22}a_{33} + a_{12}a_{23}a_{31} + a_{13}a_{21}a_{32}) \\ &-(a_{13}a_{22}a_{31} + a_{11}a_{23}a_{32} + a_{12}a_{21}a_{33}). \end{aligned} \tag{3.1}$$

We remark that the three numbers in each of terms are from different rows and different columns of A.

If we write $|A|$ and adjoin to its the first two columns, we get

$$\begin{vmatrix} a_{11} & a_{12} & a_{13} \\ a_{21} & a_{22} & a_{23} \\ a_{31} & a_{32} & a_{33} \end{vmatrix} \begin{matrix} a_{11} & a_{12} \\ a_{21} & a_{22} \\ a_{31} & a_{32} \end{matrix} \tag{3.2}$$

The first three terms in (3.1):

$$a_{11}a_{22}a_{33}, \quad a_{12}a_{23}a_{31}, \quad a_{13}a_{21}a_{32},$$

can be obtained from (3.2) in the following way:

Drawing an arrow from a_{11} to a_{33} through a_{22} (from the top left to the bottom right), we see that the product of the three numbers a_{11}, a_{22}, a_{33} on the arrow is the first product in (3.1). Drawing an arrow from a_{12} to a_{31} through a_{23} and from a_{13} to a_{32} through a_{21}, you get other two products. Similarly, you can get the three terms

$$a_{13}a_{22}a_{31}, \quad a_{11}a_{23}a_{32}, \quad a_{12}a_{21}a_{33}$$

by drawing an arrow from a_{13} to a_{31} through a_{22} (from the top right to the bottom left), from a_{11} to a_{32} through a_{23} and from a_{12} to a_{33} through a_{21}.

Example 3.2.1. Calculate the determinant

$$|A| = \begin{vmatrix} 2 & 4 & 6 \\ -4 & 5 & 6 \\ 7 & -8 & 9 \end{vmatrix}.$$

Solution. By (3.1) and (3.2), we obtain

$$\begin{aligned} |A| &= \left|\begin{array}{ccc|cc} 2 & 4 & 6 & 2 & 4 \\ -4 & 5 & 6 & -4 & 5 \\ 7 & -8 & 9 & 7 & -8 \end{array}\right| = [(2)(5)(9) + (4)(6)(7) + (6)(-4)(-8)] \\ &\quad -[(6)(5)(7) + (2)(6)(-8) + (4)(-4)(9)] = 480. \end{aligned}$$

□

Remark 3.2.2. The method given above does not work for a matrix of order greater than 3.

Before we introduce the determinants for matrices with order greater than 3, we can reorganize the six terms in (3.1) and give another definition.

$$\begin{aligned} |A| &= \begin{vmatrix} a_{11} & a_{12} & a_{13} \\ a_{21} & a_{22} & a_{23} \\ a_{31} & a_{32} & a_{33} \end{vmatrix} = (a_{11}a_{22}a_{33} + a_{12}a_{23}a_{31} + a_{13}a_{21}a_{32}) \\ &\quad -(a_{13}a_{22}a_{31} + a_{11}a_{23}a_{32} + a_{12}a_{21}a_{33}) \\ &= a_{11}(a_{22}a_{33} - a_{23}a_{32}) - a_{12}(a_{21}a_{33} - a_{23}a_{31}) + a_{13}(a_{21}a_{32} - a_{22}a_{31}) \\ &= a_{11}\begin{vmatrix} a_{22} & a_{23} \\ a_{32} & a_{33} \end{vmatrix} - a_{12}\begin{vmatrix} a_{21} & a_{23} \\ a_{31} & a_{33} \end{vmatrix} + a_{13}\begin{vmatrix} a_{21} & a_{22} \\ a_{31} & a_{32} \end{vmatrix} \\ &= a_{11}M_{11} - a_{12}M_{12} + a_{13}M_{13}, \end{aligned}$$

where

$$M_{11} = \begin{vmatrix} a_{22} & a_{23} \\ a_{32} & a_{33} \end{vmatrix}, \quad M_{12} = \begin{vmatrix} a_{21} & a_{23} \\ a_{31} & a_{33} \end{vmatrix}, \quad M_{13} = \begin{vmatrix} a_{21} & a_{22} \\ a_{31} & a_{32} \end{vmatrix}. \tag{3.3}$$

Remark 3.2.3. M_{11} is the determinant of the matrix obtained from A by deleting row 1 and column 1 of A, M_{12} is the determinant of the matrix obtained from A by deleting row 1 and column 2 of A and M_{13} is the determinant of the matrix obtained from A by deleting row 1 and column 3 of A.

Definition 3.2.4. M_{1i} is called the minor of a_{1i} for each $i = 1, 2, 3$.

Hence, we can compute the determinant $|A|$ by using the minors:

$$|A| = a_{11}M_{11} - a_{12}M_{12} + a_{13}M_{13}. \tag{3.4}$$

Note that the signs in (3.4) change and follow the following rule:

$$(-1)^{1+1}, \quad (-1)^{1+2}, \quad (-1)^{1+3}.$$

These powers are from subscripts of a_{11}, a_{12} and a_{13}, respectively.

We define

$$A_{11} = (-1)^{1+1}M_{11}, \quad A_{12} = (-1)^{1+2}M_{12} \quad \text{and } A_{13} = (-1)^{1+3}M_{13}. \tag{3.5}$$

Definition 3.2.5. A_{1i} is called the cofactor of a_{1i} for each $i = 1, 2, 3$.

Now, we can rewrite $|A|$ by using the cofactors:

$$|A| = a_{11}A_{11} + a_{12}A_{12} + a_{13}A_{13}. \tag{3.6}$$

Example 3.2.6. Let

$$|A| = \begin{vmatrix} 2 & 4 & 6 \\ -4 & 5 & 6 \\ 7 & -8 & 9 \end{vmatrix}.$$

(1) Compute $|A|$ by using minors.
(2) Compute $|A|$ by using cofactors.

Solution. (1) By (3.3), we have

$$\begin{aligned} M_{11} &= \begin{vmatrix} 5 & 6 \\ -8 & 9 \end{vmatrix} = (5)(9) - (6)(-8) = 93, \\ M_{12} &= \begin{vmatrix} -4 & 6 \\ 7 & 9 \end{vmatrix} = (-4)(9) - (6)(7) = -78. \\ M_{13} &= \begin{vmatrix} -4 & 5 \\ 7 & -8 \end{vmatrix} = (-4)(-8) - (5)(7) = -3. \end{aligned}$$

Using (3.4), we get

$$|A| = 2M_{11} - 4M_{12} + 6M_{13} = 2(93) - 4(-78) + 6(-3) = 480.$$

(2) By (3.5), we obtain

$$\begin{aligned} A_{11} &= (-1)^{1+1}M_{11} = M_{11} = 93. \\ A_{12} &= (-1)^{1+2}M_{12} = -M_{12} = -(-78) = 78. \\ A_{13} &= (-1)^{1+3}M_{11} = M_{13} = -3. \end{aligned}$$

It follows from (3.6) that

$$|A| = 2A_{11} + 4A_{12} + 6A_{13} = 2(93) + 4(78) + 6(-3) = 480.$$

□

In the above, we use the minors and cofactors corresponding to the first row of A. Similarly, we can introduce the minors and cofactors for the second row or the third row to find the determinants.

Theorem 3.2.7.

$$\begin{aligned}|A| &= a_{11}A_{11} + a_{12}A_{12} + a_{13}A_{13}\\ &= a_{21}A_{21} + a_{22}A_{22} + a_{23}A_{23}\\ &= a_{31}A_{31} + a_{32}A_{32} + a_{33}A_{33}.\end{aligned}$$

Example 3.2.8. Let

$$A = \begin{pmatrix} 2 & 4 & 6 \\ -4 & 5 & 6 \\ 7 & -8 & 9 \end{pmatrix}.$$

(1) Find $M_{21}, M_{22}, M_{23}, A_{21}, A_{22}, A_{23}$ and compute $|A|$ using cofactors.
(2) Find $M_{31}, M_{32}, M_{33}, A_{31}, A_{32}, A_{33}$ and compute $|A|$ using cofactors.

Solution. (1)

$$M_{21} = \begin{vmatrix} 4 & 6 \\ -8 & 9 \end{vmatrix} = 36 + 48 = 84,\ A_{21} = (-1)^{2+1}M_{21} = -M_{21} = -84.$$

$$M_{22} = \begin{vmatrix} 2 & 6 \\ 7 & 9 \end{vmatrix} = 18 - 42 = -24,\ A_{22} = (-1)^{2+2}M_{22} = M_{22} = -24.$$

$$M_{23} = \begin{vmatrix} 2 & 4 \\ 7 & 8 \end{vmatrix} = -16 - 28 = -44,\ A_{23} = (-1)^{2+3}M_{23} = -M_{23} = 44.$$

By Theorem 3.2.7, we have

$$|A| = a_{21}A_{21} + a_{22}A_{22} + a_{23}A_{23} = (-4)(-84) + (5)(-24) + (6)(44) = 480.$$

(2)

$$M_{31} = \begin{vmatrix} 4 & 6 \\ 5 & 6 \end{vmatrix} = 24 - 30 = -6,\ A_{31} = (-1)^{3+1}M_{31} = M_{31} = -6.$$

$$M_{32} = \begin{vmatrix} 2 & 6 \\ -4 & 6 \end{vmatrix} = 12 + 24 = 36,\ A_{32} = (-1)^{3+2}M_{32} = -M_{32} = -36.$$

$$M_{33} = \begin{vmatrix} 2 & 4 \\ -4 & 5 \end{vmatrix} = 10 + 16 = 26,\ A_{33} = (-1)^{3+3}M_{33} = M_{33} = 26.$$

By Theorem 3.2.7, we have

$$|A| = a_{31}A_{31} + a_{32}A_{32} + a_{33}A_{33} = (7)(-6) + (-8)(-36) + (9)(26) = 480.$$

□

Exercises

1. Calculate each of the following determinants by using (3.1) and (3.2).

$$|A| = \begin{vmatrix} 1 & 2 & 3 \\ -2 & 1 & 2 \\ 3 & -1 & 4 \end{vmatrix} \quad |B| = \begin{vmatrix} 2 & -2 & 4 \\ 4 & 3 & 1 \\ 0 & 1 & 2 \end{vmatrix} \text{ and } |C| = \begin{vmatrix} 0 & -1 & 3 \\ 4 & 1 & 2 \\ 0 & 0 & 1 \end{vmatrix}$$

2. Let $|A| = \begin{vmatrix} 1 & 2 & 3 \\ -2 & 1 & 2 \\ 3 & -1 & 4 \end{vmatrix}$.

   (1) Find $M_{11}, M_{12}, M_{13}, A_{11}, A_{12}, A_{13}$ and compute $|A|$ using cofactors.

   (2) Find $M_{21}, M_{22}, M_{23}, A_{21}, A_{22}, A_{23}$ and compute $|A|$ using cofactors.

   (3) Find $M_{31}, M_{32}, M_{33}, A_{31}, A_{32}, A_{33}$ and compute $|A|$ using cofactors.

3.3 Minors and cofactors of $n \times n$ determinants

Let A be a square matrix of order n given in (2.6), that is,

$$A = \begin{pmatrix} a_{11} & a_{12} & \cdots & a_{1n} \\ a_{21} & a_{22} & \cdots & a_{2n} \\ \vdots & \vdots & \ddots & \vdots \\ a_{n1} & a_{n2} & \cdots & a_{nn} \end{pmatrix}. \tag{3.7}$$

Definition 3.3.1. Let M_{ij} be the determinant of the $(n-1) \times (n-1)$ matrix obtained from A by deleting the ith row and the jth column of A. M_{ij} is called the minor of a_{ij} or the ijth minor of A.

$$A_{ij} = (-1)^{i+j} M_{ij}$$

is called the cofactor of a_{ij} or the ijth factor of A.

Example 3.3.2. Let

$$A = \begin{pmatrix} 1 & -3 & 5 & 6 \\ 2 & 4 & 0 & 3 \\ 1 & 5 & 9 & -2 \\ 4 & 0 & 2 & 7 \end{pmatrix}.$$

Find M_{32}, M_{24}, A_{32}, A_{24}.

Solution.

$$M_{32} = \begin{vmatrix} 1 & 5 & 6 \\ 2 & 0 & 3 \\ 4 & 2 & 7 \end{vmatrix} = 8, \quad A_{32} = (-1)^{3+2}M_{32} = -M_{32} = -8,$$

$$M_{24} = \begin{vmatrix} 1 & -3 & 5 \\ 1 & 5 & 9 \\ 4 & 0 & 2 \end{vmatrix} = -192, \quad A_{24} = (-1)^{2+4}M_{24} = M_{24} = -192.$$

□

Exercises

1. Find M_{32}, M_{24}, A_{32}, A_{24}, M_{41}, M_{43}, A_{41}, A_{43} if

$$A = \begin{pmatrix} 1 & -1 & 2 & 3 \\ 2 & 2 & 0 & 3 \\ 1 & 2 & 3 & -1 \\ -3 & 0 & 1 & 6 \end{pmatrix}.$$

2. Find M_{32}, M_{24}, A_{32}, A_{24}, M_{41}, M_{43}, A_{41}, A_{43} if

$$A = \begin{pmatrix} 1 & 0 & 2 & -1 \\ 2 & 0 & 0 & 1 \\ 1 & 0 & 1 & -1 \\ -1 & 0 & 1 & 2 \end{pmatrix}.$$

3.4 Determinants of $n \times n$ matrices

Definition 3.4.1. Let A be an $n \times n$ matrix given in (3.7). Then the determinant of A is defined by

$$|A| = a_{11}A_{11} + a_{12}A_{12} + \cdots + a_{1n}A_{1n}. \tag{3.8}$$

The expression of the right side of (3.8) is called an expansion of cofactors. In Defintion 3.4.1, we use the cofactors of elements of the first row of A to define $|A|$. In fact, the cofactors of elements of any rows of A can be used to define $|A|$ in a similar way. In fact, we have the following result.

Theorem 3.4.2. *For each fixed $i \in \mathcal{I}_n$,*

$$|A| = a_{i1}A_{i1} + a_{i2}A_{i2} + \cdots + a_{in}A_{in}.$$

Example 3.4.3. Calculate

$$|A| = \begin{vmatrix} 1 & 3 & 5 & 2 \\ 0 & -1 & 3 & 4 \\ 2 & 1 & 9 & 6 \\ 3 & 2 & 4 & 8 \end{vmatrix}.$$

Solution. By Theorem 3.4.2, we have

$$\begin{aligned} |A| &= a_{11}A_{11} + a_{12}A_{12} + a_{13}A_{13} + a_{14}A_{14} \\ &= a_{11}M_{11} - a_{12}M_{12} + a_{13}M_{13} - a_{14}M_{14} \\ &= M_{11} - 3M_{12} + 5M_{13} - 2M_{14} \\ &= \begin{vmatrix} -1 & 3 & 4 \\ 1 & 9 & 6 \\ 2 & 4 & 8 \end{vmatrix} - 3\begin{vmatrix} 0 & 3 & 4 \\ 2 & 9 & 6 \\ 3 & 4 & 8 \end{vmatrix} + 5\begin{vmatrix} 0 & -1 & 4 \\ 2 & 1 & 6 \\ 3 & 2 & 8 \end{vmatrix} - 2\begin{vmatrix} 0 & -1 & 3 \\ 2 & 1 & 9 \\ 3 & 2 & 4 \end{vmatrix} \\ &= -92 - 3(-70) + 5(2) - 2(-16) = 160. \end{aligned}$$

□

By Theorem 3.4.2, we can choose any of rows of A together with its cofactors to compute the determinant of A. In Example 3.4.3, we can choose row 2 or row 3 to compute $|A|$. However, the smart choice would be to choose a row which contains more zero entries. By Theorem 3.4.2, we see that we only need to compute the cofactors of the nonzero entries in that row. In Example 3.4.3, the smart choice would be to choose the row 2.

Example 3.4.4. Calculate the determinant $|A|$ given in Example 3.4.3 by using the expansion of cofactors of row 2.

Solution.

$$\begin{aligned} |A| &= a_{21}A_{21} + a_{22}A_{22} + a_{23}A_{23} + a_{24}A_{24} \\ &= 0 + (-1)A_{22} + 3A_{23} + 4A_{24} \\ &= M_{22} - 3M_{23} + 4M_{24} \\ &= \begin{vmatrix} 1 & 5 & 2 \\ 2 & 9 & 6 \\ 3 & 4 & 8 \end{vmatrix} - 3\begin{vmatrix} 1 & 3 & 2 \\ 2 & 1 & 6 \\ 3 & 2 & 8 \end{vmatrix} + 4\begin{vmatrix} 1 & 3 & 5 \\ 2 & 1 & 9 \\ 3 & 2 & 4 \end{vmatrix} \\ &= 160. \end{aligned}$$

□

The following example shows that it is important to choose a suitable row to compute a determinant.

Example 3.4.5. Evaluate

$$|A| = \begin{vmatrix} 0 & 2 & 3 & 4 \\ 0 & 1 & 0 & 1 \\ 0 & 0 & 1 & 0 \\ 3 & 1 & 2 & 0 \end{vmatrix}.$$

Solution. We use the expansion of cofactors of row 3 to compute $|A|$.

$$|A| = 0A_{31} + 0A_{32} + A_{33} + 0A_{34} = A_{33} = \begin{vmatrix} 0 & 2 & 4 \\ 0 & 1 & 1 \\ 3 & 1 & 0 \end{vmatrix} = -6.$$

□

Exercises

1. Compute the following determinants by using the expansion of cofactors of each row.

$$|A| = \begin{vmatrix} 1 & 0 & 2 & -1 \\ 2 & 0 & 0 & 1 \\ 1 & 0 & 1 & -1 \\ -1 & 0 & 1 & 2 \end{vmatrix} \qquad |B| = \begin{vmatrix} 1 & -1 & 2 & 3 \\ 2 & 2 & 0 & 3 \\ 1 & 2 & 3 & -1 \\ -3 & 0 & 1 & 6 \end{vmatrix}.$$

2. Let

$$|A| = \begin{vmatrix} 0 & 0 & 0 & 1 \\ 0 & 1 & 0 & 1 \\ 0 & 0 & 1 & 0 \\ 1 & 1 & -1 & 0 \end{vmatrix}.$$

Evaluate $|A|$ by using the expansion of cofactors of a suitable row.

3.5 Determinants of some special matrices

Theorem 3.5.1. *If A has a row of zero, then $|A| = 0$.*

Proof. Assume that all entries of ith row are zero, that is,

$$a_{i1} = a_{i2} = \cdots = a_{in} = 0.$$

Using the expansion of cofactors of ith row and Theorem 3.4.2, we have

$$|A| = a_{i1}A_{i1} + a_{i2}A_{i2} + \cdots + a_{in}A_{in} = 0A_{i1} + 0A_{i2} + \cdots + 0A_{in} = 0.$$

□

Example 3.5.2. Evaluate

$$\begin{vmatrix} 2 & 3 & 5 \\ 0 & 0 & 0 \\ 1 & -2 & 4 \end{vmatrix}.$$

Solution. Since the matrix contains a zero row, so the determinant is zero. □

Theorem 3.5.3. *If A has two equal rows, then $|A| = 0$.*

Example 3.5.4. Evaluate

$$\begin{vmatrix} 1 & 3 & -2 & 4 \\ 2 & 7 & -4 & 8 \\ 3 & 9 & 1 & 5 \\ 1 & 3 & -2 & 4 \end{vmatrix}.$$

Solution. Since the matrix has two equal rows, that is, row 1 and row 4 are same, so the determinant is zero. □

Theorem 3.5.5. *If A has two proportional rows, then $|A| = 0$.*

Example 3.5.6. Evaluate

$$\begin{vmatrix} 1 & 3 & -2 & 4 \\ 3 & 9 & 1 & 5 \\ 2 & 6 & -4 & 8 \\ 1 & 1 & 4 & 8 \end{vmatrix}.$$

Solution. Since the matrix has two proportional rows, that is, row 1 and row 3 are proportional, so the determinant is zero. □

Theorem 3.5.7. *The determinant of a triangular matrix equals the product of its diagonal entries.*

If A is an upper, a lower, or a diagonal matrix, then

$$|A| = a_{11}a_{22}\cdots a_{nn}. \tag{3.9}$$

Example 3.5.8. Evaluate the following determinants.

$$|A| = \begin{vmatrix} 2 & 1 & 7 \\ 0 & 2 & -5 \\ 0 & 0 & 1 \end{vmatrix}, \quad |B| = \begin{vmatrix} 5 & 0 & 0 & 0 \\ 2 & 3 & 0 & 0 \\ 1 & 6 & 7 & 0 \\ 0 & 0 & 2 & 1 \end{vmatrix}, \quad |C| = \begin{vmatrix} -2 & 3 & 0 & 1 \\ 0 & 0 & 2 & 4 \\ 0 & 0 & 1 & 3 \\ 0 & 0 & 0 & 2 \end{vmatrix}.$$

Solution. Since A, B, C are an upper, a lower and a diagonal matrix, respectively, by (3.9) we obtain

$$|A| = \begin{vmatrix} 2 & 1 & 7 \\ 0 & 2 & -5 \\ 0 & 0 & 1 \end{vmatrix} = 2 \times 2 \times 1 = 4.$$

$$|B| = \begin{vmatrix} 5 & 0 & 0 & 0 \\ 2 & 3 & 0 & 0 \\ 1 & 6 & 7 & 0 \\ 0 & 0 & 2 & 1 \end{vmatrix} = 5 \times 3 \times 7 \times 1 = 15 \times 7 = 105.$$

$$|C| = \begin{vmatrix} -2 & 3 & 0 & 1 \\ 0 & 0 & 2 & 4 \\ 0 & 0 & 1 & 3 \\ 0 & 0 & 0 & 2 \end{vmatrix} = (-2) \times 0 \times 1 \times 2 = 0.$$

□

Exercises

1. Evaluate each of the following determinants by inspection.

$$|A| = \begin{vmatrix} 1 & 2 & -3 \\ 0 & 1 & 2 \\ -2 & -4 & 6 \end{vmatrix} \quad |B| = \begin{vmatrix} 2 & -2 & 4 \\ 2 & -2 & 4 \\ 0 & 1 & 2 \end{vmatrix} \quad |C| = \begin{vmatrix} 1 & 2 & 3 \\ -2 & 1 & 2 \\ 1 & -\frac{1}{2} & -1 \end{vmatrix}$$

$$|D| = \begin{vmatrix} 1 & -1 & 3 & 5 & 0 \\ 0 & 0 & 0 & 0 & 0 \\ 0 & 0 & 1 & 2 & 7 \\ 5 & 6 & 3 & -4 & 6 \\ 2 & -2 & 6 & 10 & 0 \end{vmatrix} \quad |E| = \begin{vmatrix} 1 & -1 & 3 & 5 & 0 \\ 0 & 3 & 0 & 0 & 0 \\ 0 & 0 & -1 & 2 & 7 \\ 0 & 0 & 0 & -4 & 6 \\ 0 & 0 & 0 & 0 & 2 \end{vmatrix}$$

$$|F| = \begin{vmatrix} 1 & 0 & 0 & 0 \\ -2 & 2 & 0 & 0 \\ -1 & 6 & -6 & 0 \\ 4 & 0 & 2 & -1 \end{vmatrix} \quad |G| = \begin{vmatrix} -2 & 0 & 0 & 0 \\ 0 & 3 & 0 & 0 \\ 0 & 0 & -1 & 0 \\ 0 & 0 & 0 & 2 \end{vmatrix}$$

3.6 Evaluating determinants by row operations

By (3.9), we see that the determinant of an upper triangular matrix can be easily calculated. Hence, we can use row operations to change a matrix A to an upper triangular matrix B and then evaluate $|A|$ by calculating $|B|$. One can show that there exists a constant $k \in \mathbb{R}$ such that

$$|A| = k|B|.$$

In the following we show the relations between $|A|$ and $|B|$ if B is obtained from A by a row operation.

Theorem 3.6.1. (1) *If* $A \xrightarrow{R_i(c)} B$, *then* $|A| = \frac{1}{c}|B|$.

(2) *If* $A \xrightarrow{R_i(c)+R_j} B$, *then* $|A| = |B|$.

(3) *If* $A \xrightarrow{R_{i,j}} B$, *then* $|A| = -|B|$.

We can simply state Theorem 3.6.1 (1) as follows:

$$\begin{vmatrix} a_{11} & a_{12} & \cdots & a_{1n} \\ a_{21} & a_{22} & \cdots & a_{2n} \\ \vdots & \vdots & \vdots & \vdots \\ ca_{i1} & ca_{i2} & \cdots & ca_{in} \\ \vdots & \vdots & \vdots & \vdots \\ a_{n1} & a_{n2} & \cdots & a_{nn} \end{vmatrix} = c \begin{vmatrix} a_{11} & a_{12} & \cdots & a_{1n} \\ a_{21} & a_{22} & \cdots & a_{2n} \\ \vdots & \vdots & \vdots & \vdots \\ a_{i1} & a_{i2} & \cdots & a_{in} \\ \vdots & \vdots & \vdots & \vdots \\ a_{n1} & a_{n2} & \cdots & a_{nn} \end{vmatrix}. \tag{3.10}$$

In other words, we can take the common factor of one row out of the determinant.

Theorem 3.6.1 (2) means that adding the ith-row multiplied by a nonzero constant c to the jth row, the determinant keeps unchanged.

Theorem 3.6.1 (3) means that interchanging two rows, the determinant changes the sign.

Example 3.6.2. Compute

$$|A| = \begin{vmatrix} 1 & -1 & 2 \\ 12 & 4 & 16 \\ 0 & -2 & 5 \end{vmatrix}.$$

Solution.

$$\begin{aligned} |A| &= \begin{vmatrix} 1 & -1 & 2 \\ 12 & 4 & 16 \\ 0 & -2 & 5 \end{vmatrix} = \begin{vmatrix} 1 & -1 & 2 \\ (4)(3) & (4)(1) & (4)(4) \\ 0 & -2 & 5 \end{vmatrix} = 4 \begin{vmatrix} 1 & -1 & 2 \\ 3 & 1 & 4 \\ 0 & -2 & 5 \end{vmatrix} \\ &= 4(16) = 64. \end{aligned}$$

□

Example 3.6.3. Compute

$$|A| = \begin{vmatrix} 2 & -3 & 5 \\ 1 & 7 & 2 \\ -4 & 6 & -10 \end{vmatrix}.$$

Solution. $|A| = \begin{vmatrix} 2 & -3 & 5 \\ 1 & 7 & 2 \\ -4 & 6 & -10 \end{vmatrix} \overset{R_1(2)+R_3}{=\!=\!=} \begin{vmatrix} 2 & -3 & 5 \\ 1 & 7 & 2 \\ 0 & 0 & 0 \end{vmatrix} = 0.$ □

Example 3.6.4. Evaluate

$$|A| = \begin{vmatrix} 0 & 0 & 1 \\ 0 & 2 & 3 \\ 1 & -2 & 5 \end{vmatrix}.$$

Solution. $|A| = \begin{vmatrix} 0 & 0 & 1 \\ 0 & 2 & 3 \\ 1 & -2 & 5 \end{vmatrix} \overset{R_{1,3}}{=\!=} - \begin{vmatrix} 1 & -2 & 5 \\ 0 & 2 & 3 \\ 0 & 0 & 1 \end{vmatrix} = 2.$ □

In the following examples, we use row operations to calculate determinants, that is, use row operations to change A to an upper triangular matrix.

Example 3.6.5. Evaluate

$$|A| = \begin{vmatrix} 0 & 1 & 5 \\ 3 & -6 & 9 \\ 2 & 6 & 1 \end{vmatrix}.$$

Solution.

$$\begin{aligned} |A| &= \begin{vmatrix} 0 & 1 & 5 \\ 3 & -6 & 9 \\ 2 & 6 & 1 \end{vmatrix} \overset{R_{1,2}}{=\!=\!=} - \begin{vmatrix} 3 & -6 & 9 \\ 0 & 1 & 5 \\ 2 & 6 & 1 \end{vmatrix} = -3 \begin{vmatrix} 1 & -2 & 3 \\ 0 & 1 & 5 \\ 2 & 6 & 1 \end{vmatrix} \\ &\overset{R_1(-2)+R_3}{=\!=\!=\!=\!=\!=} -3 \begin{vmatrix} 1 & -2 & 3 \\ 0 & 1 & 5 \\ 0 & 10 & -5 \end{vmatrix} \overset{R_2(-10)+R_3}{=\!=\!=\!=\!=\!=} \\ &-3 \begin{vmatrix} 1 & -2 & 3 \\ 0 & 1 & 5 \\ 0 & 0 & -55 \end{vmatrix} = -3(-55) = 165. \end{aligned}$$

□

Example 3.6.6. Calculate the following determinant

$$|A| = \begin{vmatrix} 2 & 6 & 10 & 4 \\ 0 & -1 & 3 & 4 \\ 2 & 1 & 9 & 6 \\ 3 & 2 & 4 & 8 \end{vmatrix}$$

Solution.

$$|A| = \begin{vmatrix} 2 & 6 & 10 & 4 \\ 0 & -1 & 3 & 4 \\ 2 & 1 & 9 & 6 \\ 3 & 2 & 4 & 8 \end{vmatrix} = 2\begin{vmatrix} 1 & 3 & 5 & 2 \\ 0 & -1 & 3 & 4 \\ 2 & 1 & 9 & 6 \\ 3 & 2 & 4 & 8 \end{vmatrix} \overset{R_1(-2)+R_3}{\underset{R_1(-3)+R_4}{=\!=\!=}}$$

$$2\begin{vmatrix} 1 & 3 & 5 & 2 \\ 0 & -1 & 3 & 4 \\ 0 & -5 & -1 & 2 \\ 0 & -7 & -11 & 2 \end{vmatrix} \overset{R_2(-5)+R_3}{\underset{R_2(-7)+R_4}{=\!=\!=}} 2\begin{vmatrix} 1 & 3 & 5 & 2 \\ 0 & -1 & 3 & 4 \\ 0 & 0 & -16 & -18 \\ 0 & 0 & -32 & -26 \end{vmatrix}$$

$$\overset{R_3(-2)+R_4}{=\!=\!=} 2\begin{vmatrix} 1 & 3 & 5 & 2 \\ 0 & -1 & 3 & 4 \\ 0 & 0 & -16 & -18 \\ 0 & 0 & 0 & 10 \end{vmatrix} = 2[(1)(-1)(-16)(10)]$$

$$= 320.$$

□

Exercises

1. Use row operations to evaluate each of the following determinants.

$$|A| = \begin{vmatrix} 1 & 2 & -3 & -2 \\ 2 & 1 & 2 & 0 \\ -2 & -3 & 1 & -2 \\ 1 & -1 & 2 & 0 \end{vmatrix} \qquad |B| = \begin{vmatrix} 2 & 1 & 4 \\ 1 & -2 & 4 \\ 0 & 2 & 4 \end{vmatrix}$$

$$|C| = \begin{vmatrix} 0 & -1 & 3 & 5 & 0 \\ 1 & -1 & 2 & 4 & -1 \\ 0 & 0 & 1 & 2 & 7 \\ 1 & 1 & -2 & 3 & 1 \\ 2 & -2 & 6 & 10 & 0 \end{vmatrix} \qquad |D| = \begin{vmatrix} 0 & 2 & 3 \\ -2 & 1 & 2 \\ 3 & -1 & -1 \end{vmatrix}$$

$$|E| = \begin{vmatrix} 1 & -1 & 3 & 5 & 0 \\ 1 & 2 & 0 & 0 & 0 \\ -2 & 0 & -1 & 2 & 1 \\ 3 & 0 & 0 & -6 & 6 \\ 2 & 0 & 0 & 0 & 1 \end{vmatrix} \qquad |F| = \begin{vmatrix} 0 & 0 & 0 & 0 \\ -2 & 2 & 0 & 0 \\ 0 & 6 & -6 & 0 \\ 4 & 0 & 2 & -1 \end{vmatrix}$$

$$|G| = \begin{vmatrix} -2 & 0 & 0 & 0 \\ 3 & 3 & 0 & 1 \\ 0 & 1 & -1 & 0 \\ -1 & 0 & 0 & 2 \end{vmatrix}$$

3.7 Determinant of A^T

The following result shows that the determinants of A and its transpose A^T are same.

Theorem 3.7.1. $|A| = |A^T|$.

In some cases, it may be simpler to evaluate $|A|$ by computing $|A^T|$.

Example 3.7.2. Evaluate

$$|A| = \begin{vmatrix} 1 & 0 & 0 & 3 \\ 2 & 7 & 0 & 6 \\ 0 & 6 & 3 & 0 \\ 7 & 3 & 1 & 20 \end{vmatrix}.$$

Solution. $|A^T| = \begin{vmatrix} 1 & 2 & 0 & 7 \\ 0 & 7 & 6 & 3 \\ 0 & 0 & 3 & 1 \\ 3 & 6 & 0 & 20 \end{vmatrix} \underset{}{\overset{R_1(-3)+R_4}{=\!=\!=}} \begin{vmatrix} 1 & 2 & 0 & 7 \\ 0 & 7 & 6 & 3 \\ 0 & 0 & 3 & 1 \\ 0 & 0 & 0 & -1 \end{vmatrix} = -21.$ □

Exercises

1. Evaluate each of the following determinants by using its transpose.

$$|A| = \begin{vmatrix} 1 & 0 & 0 & 1 \\ 2 & 7 & 0 & 2 \\ 0 & 6 & 3 & 0 \\ 5 & 3 & 1 & 2 \end{vmatrix} \qquad |B| = \begin{vmatrix} 1 & 0 & 0 & 0 & 0 \\ 2 & 7 & 0 & 2 & 1 \\ 0 & 6 & 3 & 0 & 0 \\ 1 & 3 & 1 & 2 & -1 \\ 0 & 6 & 1 & 2 & 1 \end{vmatrix}$$

3.8 Properties of the determinants

Let A be an $n \times n$ matrix given in (3.7) and c a constant. Then

$$cA = \begin{pmatrix} ca_{11} & ca_{12} & \cdots & ca_{1n} \\ ca_{21} & ca_{22} & \cdots & ca_{2n} \\ \vdots & \vdots & \ddots & \vdots \\ ca_{n1} & ca_{n2} & \cdots & ca_{nn} \end{pmatrix}.$$

The following result gives the relation between determinants $|cA|$ and $|A|$.

Theorem 3.8.1. $|cA| = c^n|A|$.

Proof. Repeating (3.10) implies

$$
\begin{aligned}
|cA| &= \begin{vmatrix} ca_{11} & ca_{12} & \cdots & ca_{1n} \\ ca_{21} & ca_{22} & \cdots & ca_{2n} \\ \vdots & \vdots & \ddots & \vdots \\ ca_{n1} & ca_{n2} & \cdots & ca_{nn} \end{vmatrix} = c \begin{vmatrix} a_{11} & a_{12} & \cdots & a_{1n} \\ ca_{21} & ca_{22} & \cdots & ca_{2n} \\ \vdots & \vdots & \ddots & \vdots \\ ca_{n1} & ca_{n2} & \cdots & ca_{nn} \end{vmatrix} \\
&= c^2 \begin{vmatrix} a_{11} & a_{12} & \cdots & a_{1n} \\ a_{21} & a_{22} & \cdots & a_{2n} \\ \vdots & \vdots & \ddots & \vdots \\ ca_{n1} & ca_{n2} & \cdots & ca_{nn} \end{vmatrix} = \cdots = c^n \begin{vmatrix} a_{11} & a_{12} & \cdots & a_{1n} \\ a_{21} & a_{22} & \cdots & a_{2n} \\ \vdots & \vdots & \ddots & \vdots \\ a_{n1} & a_{n2} & \cdots & a_{nn} \end{vmatrix} \\
&= c^n|A|.
\end{aligned}
$$

□

Example 3.8.2. Let A be a 3×3 matrix. Assume that $|A| = 6$. Compute

$$|2A| \quad \text{and } |(3A)^T|.$$

Solution. By Theorem 3.8.1, we have

$$|2A| = 2^3|A| = 8 \times 6 = 48.$$

By Theorems 3.7.1 and 3.8.1,

$$|(3A)^T| = |3A| = 3^3|A| = 27(3) = 81.$$

□

Theorem 3.8.3. *Let A and B be $n \times n$ matrices. Then*

$$|AB| = |A||B|.$$

Example 3.8.4. Verify that $|AB| = |A||B|$ if

$$A = \begin{pmatrix} 3 & 1 \\ 2 & 1 \end{pmatrix} \quad \text{and } B = \begin{pmatrix} -1 & 3 \\ 5 & 8 \end{pmatrix}.$$

Solution. Since $|A| = 3 - 2 = 1$ and $|B| = -8 - 15 = -23$, we have

$$|A||B| = 1(-23) = -23.$$

On the other hand, since

$$AB = \begin{pmatrix} 3 & 1 \\ 2 & 1 \end{pmatrix}\begin{pmatrix} -1 & 3 \\ 5 & 8 \end{pmatrix} = \begin{pmatrix} 2 & 17 \\ 3 & 14 \end{pmatrix},$$

$|AB| = \begin{vmatrix} 2 & 17 \\ 3 & 14 \end{vmatrix} = -23$. Hence, $|AB| = |A||B|$. □

Theorem 3.8.5. *If A is invertible, then $|A| \neq 0$ and $|A^{-1}| = \dfrac{1}{|A|}$.*

Proof. Since $AA^{-1} = I$ and $|I| = 1$, it follows from Theorem 3.8.3 that

$$|AA^{-1}| = |A||A^{-1}| = |I| = 1.$$

This implies that $|A| \neq 0$ and $|A^{-1}| = \dfrac{1}{|A|}$. □

Example 3.8.6. Let A be a 3×3 matrix. Assume that $|A| = 6$. Compute

$$|(2A^{-1})^T|.$$

Solution. By Theorems 3.7.1, 3.8.1 and 3.8.5, we obtain

$$|(2A^{-1})^T| = |2A^{-1}| = 2^3|A^{-1}| = 8\frac{1}{|A|} = \frac{8}{6} = \frac{4}{3}.$$

□

Theorem 3.8.7. *Let A be an $n \times n$ matrix. Then the following assertions hold.*

(i) A is invertible if and only if $|A| \neq 0$.
(ii) A is not invertible if and only if $|A| = 0$.

Example 3.8.8. Let $A = \begin{pmatrix} 1 & 2 & 3 \\ 1 & 0 & 1 \\ 2 & 4 & 6 \end{pmatrix}$. Is A invertible?

Solution. Since the first and the third rows are proportional, $|A| = 0$. By Theorem 3.8.7, A is not invertible. □

Remark 3.8.9. In general, $|A| + |B| \neq |A + B|$.

Example 3.8.10. Let $A = \begin{pmatrix} 1 & 2 \\ 2 & 5 \end{pmatrix}$ and $B = \begin{pmatrix} 3 & 1 \\ 1 & 3 \end{pmatrix}$. Verify

$$|A + B| \neq |A| + |B|.$$

Solution. Since $A + B = \begin{pmatrix} 4 & 3 \\ 3 & 8 \end{pmatrix}$, we have $|A + B| = 23$. On the other hand, $|A| = 1$, $|B| = 8$ and $|A| + |B| = 9$. Hence, $|A + B| \neq |A| + |B|$. □

However, we have the following result on the addition of two matrices.

Theorem 3.8.11. *Let*

$$A = \begin{pmatrix} a_{11} & a_{12} & \cdots & a_{1n} \\ a_{21} & a_{22} & \cdots & a_{2n} \\ \vdots & \vdots & \ddots & \vdots \\ a_{i1} & a_{i2} & \cdots & a_{in} \\ \vdots & \vdots & \ddots & \vdots \\ a_{n1} & a_{n2} & \cdots & a_{nn} \end{pmatrix} \text{ and } B = \begin{pmatrix} a_{11} & a_{12} & \cdots & a_{1n} \\ a_{21} & a_{22} & \cdots & a_{2n} \\ \vdots & \vdots & \ddots & \vdots \\ b_{i1} & b_{i2} & \cdots & b_{in} \\ \vdots & \vdots & \ddots & \vdots \\ a_{n1} & a_{n2} & \cdots & a_{nn} \end{pmatrix}.$$

Then

$$|A| + |B| = \begin{vmatrix} a_{11} & a_{12} & \cdots & a_{1n} \\ a_{21} & a_{22} & \cdots & a_{2n} \\ \vdots & \vdots & \ddots & \vdots \\ a_{i1} + b_{i1} & a_{i2} + b_{i2} & \cdots & a_{in} + b_{in} \\ \vdots & \vdots & \ddots & \vdots \\ a_{n1} & a_{n2} & \cdots & a_{nn} \end{vmatrix}.$$

We note that in Theorem 3.8.11, all rows except the ith row of the three determinants are same.

Example 3.8.12. Compute $|A + B|$ if

$$A = \begin{pmatrix} 1 & 7 & 5 \\ 2 & 0 & 3 \\ 1 & 4 & 7 \end{pmatrix} \text{ and } B = \begin{pmatrix} 1 & 7 & 5 \\ 2 & 0 & 3 \\ 0 & 1 & -1 \end{pmatrix}.$$

Solution. Since the first two rows of A and B are same, by Theorem 3.8.11, we have

$$\begin{aligned} |A + B| &= \begin{vmatrix} 1 & 7 & 5 \\ 2 & 0 & 3 \\ 1 & 4 & 7 \end{vmatrix} + \begin{vmatrix} 1 & 7 & 5 \\ 2 & 0 & 3 \\ 0 & 1 & -1 \end{vmatrix} = \begin{vmatrix} 1 & 7 & 5 \\ 2 & 0 & 3 \\ 1+0 & 4+1 & 7-1 \end{vmatrix} \\ &= \begin{vmatrix} 1 & 7 & 5 \\ 2 & 0 & 3 \\ 1 & 5 & 6 \end{vmatrix} = -28. \end{aligned}$$

□

Exercises

1. Let A be a 4×4 matrix. Assume that $|A| = 2$. Compute each of the following determinants.

   (1) $|-3A|$ (2) $|A^{-1}|$ (3) $|A^T|$ (4) $|A^3|$ (5) $|(2A^{-1})^T|$ (6) $|(2(-A)^T)^{-1}|$

2. Determine whether the given matrix is invertible.

$$A = \begin{pmatrix} 1 & 0 & 0 & 0 \\ 2 & 1 & 2 & 0 \\ -2 & -3 & 1 & -2 \\ 1 & -1 & 0 & 0 \end{pmatrix} \qquad B = \begin{pmatrix} 2 & 1 & 4 \\ 1 & -2 & 4 \\ -2 & -1 & -4 \end{pmatrix}$$

$$C = \begin{pmatrix} 0 & -1 & 3 & 5 & 2 \\ 1 & -1 & 2 & 4 & -1 \\ 0 & 0 & 1 & 2 & 7 \\ 1 & 1 & -2 & 3 & 1 \\ 0 & -2 & 6 & 10 & 4 \end{pmatrix} \qquad D = \begin{pmatrix} 0 & 1 & 3 \\ -2 & 1 & 2 \\ 1 & -1 & -1 \end{pmatrix}$$

$$E = \begin{pmatrix} 1 & -1 & 3 & 5 & 0 \\ 0 & 0 & 0 & 0 & 0 \\ -2 & 0 & -1 & 2 & 1 \\ 3 & 0 & 0 & -6 & 6 \\ 2 & 0 & 0 & 0 & 1 \end{pmatrix} \qquad F = \begin{pmatrix} 1 & 0 & 0 & 0 \\ -2 & 2 & 0 & 0 \\ 0 & 6 & -6 & 0 \\ 4 & 0 & 2 & -1 \end{pmatrix}$$

3. Find all real numbers x such that the following matrix is invertible.

$$A = \begin{pmatrix} x & 0 & 0 & 0 \\ 2 & x & 2 & 0 \\ -2 & -3 & 1 & -2 \\ 1 & 0 & 0 & 1 \end{pmatrix} \qquad B = \begin{pmatrix} 2 & x & 4 \\ 1 & -2 & 4 \\ -2 & -1 & x \end{pmatrix}$$

4. Let $A = \begin{pmatrix} 2 & 1 & 3 \\ 1 & 0 & 3 \\ 1 & 4 & 7 \end{pmatrix}$ and $B = \begin{pmatrix} 2 & 1 & 3 \\ 1 & 0 & 3 \\ 0 & 1 & -1 \end{pmatrix}$. Compute $|A + B|$.

5. Compute $|A + B|$ if

$$A = \begin{pmatrix} 1 & 0 & 0 & 2 \\ -1 & 0 & 1 & 2 \\ 0 & 1 & 4 & 3 \\ -2 & 2 & 0 & 1 \end{pmatrix} \quad \text{and } B = \begin{pmatrix} 1 & 0 & 0 & 2 \\ 3 & 0 & 0 & 1 \\ 0 & 1 & 4 & 3 \\ -2 & 2 & 0 & 1 \end{pmatrix}.$$

Chapter 4

Systems of linear equations

4.1 Linear equations with 2 variables

The equation of the form

$$ax + by = c \tag{4.1}$$

is called a linear equation, where a, b, c are constants and x, y are variables.

Remark 4.1.1. In a linear equation, the power of each variable must be 1 and there are no terms containing the product of two variables. Hence, in an equation, if there is a variable whose power is not 1 or if there is a term which contains a product of two variables, then the equation is not a linear equation.

Example 4.1.2. Determine which of the following equations are linear.
(1) $x + y = 1$. (2) $0x + 0y = 1$. (3) $0x + 0y = 0$. (4) $2x + \frac{3}{2}y = 4$.
(5) $x^2 + y^2 = 1$. (6) $xy = 1$. (7) $4x^2 + 6y^2 = 1$. (8) $4xy + 6y = 3$.
(9) $y = \sin x$. (10) $\frac{1}{\sqrt{2}}x + 10^{\frac{2}{3}}y = 2$. (11) $x^{\frac{1}{\sqrt{2}}} + 4y^{\frac{2}{3}} = 2$.

Solution. Only (1)-(4) and (10) are linear equations. □

If $x = x_0$, $y = y_0$ satisfies (4.1), that is

$$ax_0 + by_0 = c,$$

then $x = x_0$, $y = y_0$ is called a solution of (4.1).

We often write the solution in the vector form: $\vec{X} = (x_0, y_0)$ or $\vec{X} = \begin{pmatrix} x_0 \\ y_0 \end{pmatrix}$ or treat the solution as a point $P(x_0, y_0)$ in $\mathbb{R}^2$.

If either a or b is not zero, then (4.1) is an equation of a straight line. A point $P(x_0, y_0)$ is a solution of (4.1) if and only if the point $P(x_0, y_0)$ is on the line. In other words, a point $P(x_0, y_0)$ is not on the line if and only if it is not a solution of the equation.

Example 4.1.3. Consider the linear equation

$$x + y = 1.$$

Verify whether the following points are the solutions of the above linear equation:

$P_1(0,1)$, $P_2(1,0)$, $P_3(1,1)$ and $P_4(x, 1-x)$ for each $x \in \mathbb{R}$.

Solution. It is readily verified that $P_1(0,1)$, $P_2(1,0)$ are solutions and $P_3(1,1)$ is not a solution. We see that $P_4(x, 1-x)$ is a solution for each $x \in \mathbb{R}$ since $x + (1-x) = 1$. $\square$

Since $(x, 1-x)$ is a solution for each $x \in \mathbb{R}$, the equation $x + y = 1$ has infinite many solutions. In geometry, every point on the line is a solution.

Solving a linear equation means finding all the solutions of the linear equation.

Example 4.1.4. Solve the linear equation

$$x + y = 1$$

and express the solution by a linear combination of two vectors.

Solution. **(Method 1)** Let $x = t$. Then $y = 1 - x = 1 - t$. Hence, $\begin{pmatrix} x \\ y \end{pmatrix} = \begin{pmatrix} 0+t \\ 1-t \end{pmatrix} = \begin{pmatrix} t \\ 1-t \end{pmatrix}$ is a solution for each $t \in \mathbb{R}$.

We can rewrite the solution into a linear combination of two vectors as follows:

$$\begin{pmatrix} x \\ y \end{pmatrix} = \begin{pmatrix} 0 \\ 1 \end{pmatrix} + t \begin{pmatrix} 1 \\ -1 \end{pmatrix}.$$

(Method 2) Let $y = s$. Then $x = 1 - y = 1 - s$. Hence,

$$\begin{pmatrix} x \\ y \end{pmatrix} = \begin{pmatrix} 1-s \\ s \end{pmatrix} = \begin{pmatrix} 1-s \\ 0+s \end{pmatrix} = \begin{pmatrix} 1 \\ 0 \end{pmatrix} + s \begin{pmatrix} -1 \\ 1 \end{pmatrix}$$

is a solution for each $s \in \mathbb{R}$. $\square$

Example 4.1.5. Solve the linear equation

$$0x + 0y = 2.$$

Solution. For each $(x, y) \in \mathbb{R}^2$, $0x + 0y = 0 \neq 2$, that is, (x, y) does not satisfy the equation. Hence, the linear equation has no solutions. $\square$

Example 4.1.6. Consider the linear equation

$$4x - 2y = 1.$$

Verify whether the following points:

$$(0, -\frac{1}{2}), \quad (0, \frac{1}{2}), \quad (1, 1)$$

are solutions of the linear equation and solve this equation.

Solution. $(0, -\frac{1}{2})$ is a solution since $4(0) - 2(-\frac{1}{2}) = 1$; $(0, \frac{1}{2})$ is not a solution since $4(0) - 2(\frac{1}{2}) = -1 \neq 1$; $(1, 1)$ is not a solution since $4(1) - 2(1) = 2 \neq 1$.

Let $y = t$. Then $4x - 2t = 1$, $4x = 1 + 2t$ and $x = \frac{1}{4} + \frac{1}{2}t$. The solution to the linear equation is

$$x = \frac{1}{4} + \frac{1}{2}t \quad \text{and } y = t \text{ for each } t \in (-\infty, \infty).$$

□

Exercises

1. Determine which of the following equations are linear.

   a) $2x+y=1$. *b)* $0x+0y=3$. *c)* $0x+0y=0$. *d)* $x+\frac{2}{3}y=5$.
   e) $x^3+y^4=5$. *f)* $xy=6$. *g)* $3x^2+5y^2=1$. *h)* $4xy+5y=3x$.
   i) $y=\cos x$. *j)* $\frac{1}{\sqrt{3}}x+10^{\frac{1}{3}}y=4$. *k)* $x^{\frac{1}{\sqrt{2}}}+6y^{\frac{1}{4}}=1$.
   l) $\frac{3}{\sqrt{4}}x=7^{\frac{1}{3}}x-2y+2$.

2. Consider the linear equation

   $$2x + y = 5.$$

   Verify whether the following points are the solutions of the above linear equation:

   $P_1(0,1)$, $P_2(1,0)$, $P_3(0,5)$ and $P_4(\frac{1}{2}x, 5-x)$ for each $x \in \mathbb{R}$.

3. Solve the linear equation

   $$x + y = 2$$

   and express the solution by a linear combination of two vectors.

4. Solve the linear equation $0x + 0y = 1$.

5. Verify whether the following points: $(1, -2)$, $(0, \frac{1}{2})$, $(1, 1)$ are solutions of the linear equation

   $$x - 2y = 1$$

   and solve this equation.

4.2 A system of linear equations with two variables

A set of two linear equations in two variables has the form of

$$\begin{cases} a_{11}x + a_{12}y = b_1 \\ a_{21}x + a_{22}y = b_2, \end{cases} \tag{4.2}$$

where a_{11}, a_{12}, a_{21}, a_{22}, b_1 and b_2 are constants and x, y are variables. Sometimes, we replace x and y by x_1 and x_2, respectively. In this case, (4.2) becomes

$$\begin{cases} a_{11}x_1 + a_{12}x_2 = b_1 \\ a_{21}x_1 + a_{22}x_2 = b_2. \end{cases} \tag{4.3}$$

If $x = x_0$, $y = y_0$ satisfies (4.2), that is

$$\begin{cases} a_{11}x_0 + a_{12}y_0 = b_1 \\ a_{21}x_0 + a_{22}y_0 = b_2, \end{cases}$$

then $x = x_0$, $y = y_0$ is called a solution of (4.2). Simply speaking, (x_0, y_0) is a solution of the system if it satisfies both equations, that is (x_0, y_0) is the intersection point of the two lines. (x_0, y_0) is not a solution of the system if it does not satisfy one of the two equations.

Example 4.2.1. Verify whether each of the following points:

$$P_1(6, -1), P_2(8, 1), P_3(2, 3), P_4(0, 0)$$

is a solution of system of the linear equations

$$\begin{cases} x - y = 7 ------(1) \\ x + y = 5. ------(2) \end{cases}$$

Solution. $P_1(6, -1)$ satisfies both equations, so it is a solution of the system. $P_2(8, 1)$ satisfies (1) but not (2), so it is not a solution of the system. $P_3(2, 3)$ satisfies (2) but not (1), so it is not a solution of the system. $P_4(0, 0)$ satisfies neither (1) nor (2), hence it is not a solution of the system. □

Example 4.2.2. Solve the following system.

$$\begin{cases} x_1 - x_2 = 7 ------(1) \\ x_1 + x_2 = 5. ------(2) \end{cases}$$

Solution. (1) + (2) implies $2x_1 = 12$ and $x_1 = 6$. By (1), we have $x_2 = x_1 - 7 = 6 - 7 = -1$, so $(x_1, x_2) = (6, -1)$ is a solution of the system. □

$(x_1, x_2) = (6, -1)$ is in fact the unique solution of the system. In geometry, (1) and (2) are two lines. The solution $(x_1, x_2) = (6, -1)$ is on both line (1) and line (2), so it is the intersection point of the two lines.

Example 4.2.3. Solve the following system

$$\begin{cases} x_1 + x_2 = 1 - - - - - - - - (1) \\ 2x_1 + 2x_2 = 2. - - - - - - -(2) \end{cases}$$

and express the solution by a linear combination.

Solution. The two equations are same. Hence, the solutions of (1) are the solutions of (2). Let $x_2 = t$. Then $x_1 = 1 - t$. So

$$\begin{pmatrix} x_1 \\ x_2 \end{pmatrix} = \begin{pmatrix} 1 \\ 0 \end{pmatrix} + t \begin{pmatrix} -1 \\ 1 \end{pmatrix}$$

is a solution of the system for each $t \in \mathbb{R}$. □

In geometry, (1) and (2) are a same line. The solution $(x_1, x_2) = (t, 1 - t)$ is on both line (1) and line (2), so every point on the line is a solution of the system. Hence, the system has infinite many solutions.

Example 4.2.4. Solve the following system

$$\begin{cases} x_1 + x_2 = 1 - - - - - - -(1) \\ x_1 + x_2 = 2. - - - - - - -(2) \end{cases}$$

Solution. The system has no solutions. This is because, if (x_1, x_2) is a solution of (1)-(2), (x_1, x_2) satisfies both (1) and (2). Hence, (1)−(2) implies

$$(x_1 + x_2) - (x_1 + x_2) = 1 - 2 = -1$$

and $0 = -1$, a contradiction. □

In geometry, (1) and (2) are two parallel lines, so they have no intersection points, that is, the system has no solutions.

The above examples show that a system of two linear equations with two variables has only one solution, or has infinitely many solutions or has no solutions. The same fact holds for a system which contains more than two linear equations.

One can prove that the following result holds.

Theorem 4.2.5. *A system of linear equations of two variables has only one solution, or has infinitely many solutions or has no solutions.*

Hence, if a system has two different solutions, then it follows from Theorem 4.2.5 that it has infinite many solutions.

Exercises

1. Verify which of the following points:

$$P_1(6, -\frac{1}{2}),\ P_2(8, 1),\ P_3(2, 3),\ P_4(0, 0)$$

are solutions of system of the linear equations

$$\begin{cases} x - 2y = 7 \\ x + 2y = 5. \end{cases}$$

2. Solve each of the following systems

$$(1) \begin{cases} x_1 - x_2 = 6 \\ x_1 + x_2 = 5 \end{cases} \qquad (2) \begin{cases} x_1 + 2x_2 = 1 \\ 2x_1 + x_2 = 2 \end{cases} \qquad (3) \begin{cases} x_1 + 2x_2 = 1 \\ x_1 + 2x_2 = 2. \end{cases}$$

4.3 A system of linear equations of n variables

The equation of the form

$$a_1x_1 + a_2x_2 + \cdots + a_nx_n = b. \tag{4.4}$$

is called a linear equation, where $a_1, a_2, \cdots, a_n, b$ are constants and $x_1, \cdots, x_n$ are variables.

Remark 4.3.1. In (4.4), the power of each variable is 1 and there are no terms containing the product of two or more variables.

Example 4.3.2. Determine which of the following equations are linear.
(1) $x + y + x = 1$. (2) $4x + 6y + z = 1$. (3) $\sqrt{2}x + \frac{1}{6^{\frac{2}{3}}}y = 1 - 6z$.
(4) $x_1 + x_2 + x_3 + 5x_4 = 5$. (5) $4xy + 6yz + z = 1$.

Solution. (1)-(4) are linear equations while (5) is not a linear equation. □

Let $m, n \in \mathbb{N}$. A set of m linear equations in the same n variables is called a system of linear equations. Hence, a system of linear equations is of the form:

$$\begin{cases} a_{11}x_1 + a_{12}x_2 + \cdots + a_{1n}x_n = b_1 \\ a_{21}x_1 + a_{22}x_2 + \cdots + a_{2n}x_n = b_2 \\ \vdots \\ a_{m1}x_1 + a_{m2}x_2 + \cdots + a_{mn}x_n = b_m. \end{cases} \tag{4.5}$$

The matrix

$$A = \begin{pmatrix} a_{11} & a_{12} & \cdots & a_{1n} \\ a_{21} & a_{22} & \cdots & a_{2n} \\ \vdots & \vdots & \ddots & \vdots \\ a_{m1} & a_{m2} & \cdots & a_{mn} \end{pmatrix} \tag{4.6}$$

is called the coefficient matrix of (4.5).

Let $\overrightarrow{b} = \begin{pmatrix} b_1 \\ b_2 \\ \vdots \\ b_m \end{pmatrix}$. The matrix

$$(A|\overrightarrow{b}) = \left(\begin{array}{cccc|c} a_{11} & a_{12} & \cdots & a_{1n} & b_1 \\ a_{21} & a_{22} & \cdots & a_{2n} & b_2 \\ \vdots & \vdots & \ddots & \vdots & \vdots \\ a_{m1} & a_{m2} & \cdots & a_{mn} & b_m \end{array}\right) \tag{4.7}$$

is called the augmented matrix of (4.5).

(4.5) with $\overrightarrow{b} = \overrightarrow{0}$ is called a homogeneous system of linear equations, that is,

$$\begin{cases} a_{11}x_1 + a_{12}x_2 + \cdots + a_{1n}x_n = 0 \\ a_{21}x_1 + a_{22}x_2 + \cdots + a_{2n}x_n = 0 \\ \vdots \\ a_{m1}x_1 + a_{m2}x_2 + \cdots + a_{mn}x_n = 0. \end{cases} \tag{4.8}$$

Example 4.3.3. Find the coefficient matrix and the augmented matrix of each of the following system of linear equations

(1) $x + y = 1$. (2) $\begin{cases} x_1 - x_2 = 7 \\ x_1 + x_2 = 5. \end{cases}$ (3) $\begin{cases} x_1 + x_2 + 3x_3 = 9 \\ 2x_1 + 4x_2 - 3x_3 = 1 \\ 3x_1 + 6x_2 - 5x_3 = 0. \end{cases}$

(4) $\begin{cases} 2x_1 + 3x_3 - 2 = 0 \\ 4x_2 - 3x_3 - 1 = 0 \\ x_1 + 2x_2 - x_3 = 0. \end{cases}$

Solution. (1) $A = (1\,1)$ and $(A|\overrightarrow{b}) = (1\,1|\,1)$.

(2) $A = \begin{pmatrix} 1 & -1 \\ 1 & 1 \end{pmatrix}$ and $(A|\overrightarrow{b}) = \left(\begin{array}{cc|c} 1 & -1 & 7 \\ 1 & 1 & 5 \end{array}\right)$.

(3) $A = \begin{pmatrix} 1 & 1 & 3 \\ 2 & 4 & -3 \\ 3 & 6 & -5 \end{pmatrix}$ and $(A|\overrightarrow{b}) = \left(\begin{array}{ccc|c} 1 & 1 & 3 & 9 \\ 2 & 4 & -3 & 1 \\ 3 & 6 & -5 & 0 \end{array}\right)$.

(4) We note that the entries of the first column of A are the coefficients of x_1, the entries of the second column of A is the coefficients of x_2, and so on. Hence, we rewrite the system as follows:

$$\begin{cases} 2x_1 + 0x_2 + 3x_3 = 2 \\ 0x_1 + 4x_2 - 3x_1 = 1 \\ x_1 + 2x_2 - x_3 = 0. \end{cases}$$

From the above system, the coefficient matrix and the augmented matrix are

$$A = \begin{pmatrix} 2 & 0 & 3 \\ 0 & 4 & -3 \\ 1 & 2 & -1 \end{pmatrix} \quad \text{and } (A|\overrightarrow{b}) = \left(\begin{array}{ccc|c} 2 & 0 & 3 & 2 \\ 0 & 4 & -3 & 1 \\ 1 & 2 & -1 & 0 \end{array}\right).$$

□

In many cases, we need to write out the systems corresponding to the augmented matrices.

Example 4.3.4. Find the system of linear equations corresponding to each of the augmented matrices:

$$(1) \quad \left(\begin{array}{ccc|c} 1 & 0 & 0 & 3 \\ 0 & 1 & 0 & 2 \\ 0 & 0 & 1 & 1 \end{array}\right) \quad (2) \quad \left(\begin{array}{cc|c} 2 & 0 & 0 \\ 3 & 1 & 2 \\ 0 & 1 & 1 \end{array}\right).$$

Solution. (1) The system corresponding to the augmented matrix is

$$\begin{cases} x_1 = 3 \\ x_2 = 2 \\ x_3 = 1. \end{cases}$$

(2) The system corresponding to the augmented matrix is

$$\begin{cases} 2x_1 + 0x_2 = 0 \\ 3x_1 + x_2 = 2 \\ 0x_1 + x_2 = 1 \end{cases} \quad \text{or} \quad \begin{cases} 2x_1 = 0 \\ 3x_1 + x_2 = 2 \\ x_2 = 1. \end{cases}$$

□

It is useful to write (4.5) into other expressions.

Let $\overrightarrow{X} = \begin{pmatrix} x_1 \\ x_2 \\ \vdots \\ x_n \end{pmatrix} \in \mathbb{R}^n$. Then (4.5) can be written as follows:

$$A\overrightarrow{X} = \overrightarrow{b} \tag{4.9}$$

Let $\overrightarrow{a_1}$, $\overrightarrow{a_2}$, $\cdots$, $\overrightarrow{a_n}$ are the column vectors of A given in (2.2), that is,

$$\overrightarrow{a_1} = \begin{pmatrix} a_{11} \\ a_{21} \\ \vdots \\ a_{m1} \end{pmatrix}, \quad \overrightarrow{a_2} = \begin{pmatrix} a_{12} \\ a_{22} \\ \vdots \\ a_{m2} \end{pmatrix}, \cdots, \overrightarrow{a_n} = \begin{pmatrix} a_{1n} \\ a_{2n} \\ \vdots \\ a_{mn} \end{pmatrix}.$$

Then (4.5) can be written into the following form:

$$x_1\overrightarrow{a_1} + x_2\overrightarrow{a_2} + \cdots + x_n\overrightarrow{a_n} = \overrightarrow{b}. \tag{4.10}$$

Example 4.3.5. Consider the following system of linear equations

$$\begin{cases} x_1 + x_2 + 3x_3 = 2 \\ 2x_1 + 4x_2 - 3x_3 = 1 \\ 3x_1 + 6x_2 + 5x_3 = 0. \end{cases} \tag{4.11}$$

Write the above system into the forms (4.9) and (4.10).

Solution. Let

$$A = \begin{pmatrix} 1 & 1 & 3 \\ 2 & 4 & -3 \\ 3 & 6 & 5 \end{pmatrix}, \quad \overrightarrow{X} = \begin{pmatrix} x_1 \\ x_2 \\ x_3 \end{pmatrix} \quad \text{and } \overrightarrow{b} = \begin{pmatrix} 2 \\ 1 \\ 0 \end{pmatrix}.$$

Then the system can be rewritten as $A\overrightarrow{X} = \overrightarrow{b}$.

Let

$$\overrightarrow{a_1} = \begin{pmatrix} 1 \\ 2 \\ 3 \end{pmatrix}, \quad \overrightarrow{a_2} = \begin{pmatrix} 1 \\ 4 \\ 6 \end{pmatrix}, \quad \overrightarrow{a_3} = \begin{pmatrix} 3 \\ -3 \\ 5 \end{pmatrix} \quad \text{and } \overrightarrow{b} = \begin{pmatrix} 2 \\ 1 \\ 0 \end{pmatrix}.$$

Then the system can be rewritten as

$$x_1\overrightarrow{a_1} + x_2\overrightarrow{a_2} + x_3\overrightarrow{a_3} = \overrightarrow{b}.$$

□

Example 4.3.6. Write the following system into forms (4.9) and (4.10).

$$\begin{cases} x_1 + x_2 + 3x_3 = 0 \\ 2x_1 + 4x_2 - 3x_3 = 0 \\ 3x_1 + 6x_2 - 5x_3 = 0. \end{cases}$$

Solution. Let $A = \begin{pmatrix} 1 & 1 & 3 \\ 2 & 4 & -3 \\ 3 & 6 & -5 \end{pmatrix}$, $\overrightarrow{X} = \begin{pmatrix} x_1 \\ x_2 \\ x_3 \end{pmatrix}$ and $\overrightarrow{b} = \begin{pmatrix} 0 \\ 0 \\ 0 \end{pmatrix}$. Then the system can be rewritten as

$$A\overrightarrow{X} = \overrightarrow{b}.$$

Let $\overrightarrow{a_1} = \begin{pmatrix} 1 \\ 2 \\ 3 \end{pmatrix}$, $\overrightarrow{a_2} = \begin{pmatrix} 1 \\ 4 \\ 6 \end{pmatrix}$ and $\overrightarrow{a_3} = \begin{pmatrix} 3 \\ -3 \\ -5 \end{pmatrix}$. Then the system can be rewritten as

$$x_1\overrightarrow{a_1} + x_2\overrightarrow{a_2} + x_3\overrightarrow{a_3} = \overrightarrow{0}.$$

□

If $x_1 = s_1, x_2 = s_2, \cdots, x_n = s_n$ satisfies each of the linear equations of (4.5), that is,

$$\begin{cases} a_{11}s_1 + a_{12}s_2 + \cdots + a_{1n}s_n = b_1 \\ a_{21}s_1 + a_{22}s_2 + \cdots + a_{2n}s_n = b_2 \\ \vdots \\ a_{m1}s_1 + a_{m2}s_2 + \cdots + a_{mn}s_n = b_m, \end{cases} \tag{4.12}$$

then $x_1 = s_1, x_2 = s_2, \cdots, x_n = s_n$ (or the vector $\overrightarrow{X} = (s_1, s_2, \cdots, s_n)$ or the point $P(s_1, s_2, \cdots, s_n)$) is a solution of (4.5).

Definition 4.3.7. A system of linear equations is said to be consistent if it has at least one solution. It is said to be inconsistent if it has no solutions.

By (4.10), we see that $\overrightarrow{X} = (s_1, s_2, \cdots s_n)$ is a solution of (4.5) if and only if the vector $\overrightarrow{b}$ is a linear combination of the column vectors of the coefficient matrix A. Equivalently, (4.5) has no solutions if and only if the vector $\overrightarrow{b}$ is not a linear combination of the column vectors of the coefficient matrix A.

Example 4.3.8. Consider the system of linear equations (4.11).

(1) Verify that $P_1(1, 3, -1)$ is not a solution of (4.11). $P_2(\frac{89}{19}, -\frac{42}{19}, -\frac{3}{19})$ is a solution of (4.11).

(2) Determine if the vector $\overrightarrow{b} = \begin{pmatrix} 2 \\ 1 \\ 0 \end{pmatrix}$ is a linear combination of the column vectors of the coefficient matrix of the system.

Solution. (1) It is easily verify that $P_1(1, 3, 0 - 1)$ is not a solution of (4.11) while $P_2(\frac{89}{19}, -\frac{42}{19}, -\frac{3}{19})$ is a solution of (4.11).

(2) Since $P_2(\frac{89}{19}, -\frac{42}{19}, -\frac{3}{19})$ is a solution of (4.11), $\overrightarrow{b} = \begin{pmatrix} 2 \\ 1 \\ 0 \end{pmatrix}$ is a linear combination of the column vectors of the coefficient matrix of (4.11). $\square$

It is obvious that $(x_1, x_2, \cdots, x_n) = (0, 0, \cdots 0)$ is always a solution of the homogeneous system (4.8). The solution $(0, 0, \cdots, 0)$ is called a zero solution or trivial solution of (4.8). Hence, we have the following theorem.

Theorem 4.3.9. *A homogeneous system of linear equations is consistent.*

Exercises

1. Determine which of the following equations are linear.

   a) $x + 2y + z = 6$. b) $2x - 6y + z = 1$. c) $-\sqrt{2}x + \frac{1}{6^{\frac{2}{3}}}y = 4 - 3z$.

   d) $3x_1 + 2x_2 + 4x_3 + 5x_4 = 1$. e) $2xy + 3yz + 5z = 8$.

2. Find the coefficient matrix and the augmented matrix of each of the following systems of linear equations.

   (1) $x + 2y = 6$. (2) $\begin{cases} 2x_1 - x_2 = 6 \\ 4x_1 + x_2 = 3. \end{cases}$ (3) $\begin{cases} -x_1 + 2x_2 + 3x_3 = 4 \\ 3x_1 + 2x_2 - 3x_3 = 5 \\ 2x_1 + 3x_2 - x_3 = 1. \end{cases}$

   (4) $\begin{cases} x_1 - 3x_3 - 4 = 0 \\ 2x_2 - 5x_3 - 8 = 0 \\ 3x_1 + 2x_2 - x_3 = 4. \end{cases}$

3. Find the system of linear equations corresponding to each of the following augmented matrices:

$$\left(\begin{array}{ccc|c} 1 & 2 & 3 & 1 \\ -1 & 1 & 0 & -2 \\ 0 & -1 & 1 & 0 \end{array}\right) \quad \left(\begin{array}{cc|c} 2 & 1 & 1 \\ 0 & -1 & -2 \\ 0 & 3 & -1 \end{array}\right).$$

4. Consider each of the following systems of linear equations

$$a) \begin{cases} -x_1 + x_2 + 2x_3 = 3 \\ 2x_1 + 6x_2 - 5x_3 = 2 \\ -3x_1 + 7x_2 - 5x_3 = -1. \end{cases} \qquad b) \begin{cases} x_1 - x_2 + 4x_3 = 0 \\ -2x_1 + 4x_2 - 3x_3 = 0 \\ 3x_1 + 6x_2 - 8x_3 = 0. \end{cases}$$

   Write the above system into the form $A\overrightarrow{X} = \overrightarrow{b}$ and express $\overrightarrow{b}$ as a linear combination of the column vectors of the coefficient matrix A.

5. Consider the following system of linear equations

$$\begin{cases} x_1 + x_2 + 3x_3 = 2 \\ 2x_1 + 4x_2 - 3x_3 = 1 \\ 3x_1 + 6x_2 + 5x_3 = 0. \end{cases}$$

(1) Verify that $P_1(1, 3, -1)$ is not a solution of the system. $P_2(\frac{5}{6}, \frac{7}{6}, \frac{4}{3})$ is a solution of the system.

(2) Determine if the vector $\overrightarrow{b} = \begin{pmatrix} 3 \\ 2 \\ -1 \end{pmatrix}$ is a linear combination of the column vectors of the coefficient matrix of the system.

4.4 Back-substitution

In this section, we shall solve a system of linear equations whose augmented matrix is a row echelon matrix. Before we do that, we introduce the following notations.

Basic variables and free variables The variables corresponding to the leading entries of the row echelon matrix of the coefficient matrix are called basic variables of the system and other variables are called free variables of the system.

When we solve a system of linear equations whose augmented matrix is a row echelon matrix, we will solve each linear equation for the basic variable and treat the free variables as constants. The method to be used to solve a system of linear equations whose augmented matrix is a row echelon matrix is called the back-substitution.

Example 4.4.1. Consider the system of the following linear equations

$$\begin{cases} x_1 + 2x_2 - 3x_3 = 4 ------(1) \\ 2x_2 - 6x_3 = 5 --------(2) \\ -x_3 = 2. ----------(3) \end{cases}$$

(a) Find the coefficient matrix and the augmented matrix of the system.
(b) Find the basic variables and free variables of the system.
(c) Solve the system.
(d) Determine if the vector $\overrightarrow{b} = \begin{pmatrix} 4 \\ 5 \\ 2 \end{pmatrix}$ is a linear combination of the column vectors of the coefficient matrix of the system.

Solution. (*a*) The coefficient matrix and the augmented matrix are

$$A = \begin{pmatrix} 1 & 2 & -3 \\ 0 & 2 & -6 \\ 0 & 0 & -1 \end{pmatrix} \quad \text{and } (A|\overrightarrow{b}) = \left(\begin{array}{ccc|c} 1 & 2 & -3 & 4 \\ 0 & 2 & -6 & 5 \\ 0 & 0 & -1 & 2 \end{array}\right).$$

(*b*) The coefficient matrix is a row echelon matrix and its leading entries are 1, 2 and -1. The variables corresponding to the leading entries are x_1, x_2 and x_3, respectively and there are no free variables.

(*c*) By (3), we have $x_3 = -2$. By (2), we have

$$x_2 = \frac{1}{2}(5 + 6x_3) = \frac{1}{2}[5 + 6(-2)] = -\frac{7}{2}.$$

By (1), we obtain

$$x_1 = 4 - 2x_2 + 3x_3 = 4 - 2(-\frac{7}{2}) + 3(-2) = -9.$$

Hence, $(x_1, x_2, x_3) = (-1, -\frac{7}{2}, -9)$ is a solution of the system.

(*d*) Since the system has a solution, so the vector $\overrightarrow{b} = \begin{pmatrix} 4 \\ 5 \\ 2 \end{pmatrix}$ is a linear combination of the column vectors of the coefficient matrix A, namely,

$$\overrightarrow{b} = \begin{pmatrix} 4 \\ 5 \\ 2 \end{pmatrix} = -a_1 - \frac{7}{2}a_2 - 9a_3,$$

where $\overrightarrow{a_1} = \begin{pmatrix} 1 \\ 0 \\ 0 \end{pmatrix}$, $\overrightarrow{a_2} = \begin{pmatrix} 2 \\ 2 \\ 0 \end{pmatrix}$ and $\overrightarrow{a_3} = \begin{pmatrix} -3 \\ -6 \\ -1 \end{pmatrix}$. □

Remark 4.4.2. Example 4.4.1 shows that if the leading entries are equal to the numbers of variables, the system has a unique solution.

Example 4.4.3. Consider the following system

$$\begin{cases} x_1 - x_2 - 3x_3 + x_4 = 1 - - - - - - -(1) \\ -x_2 - x_3 + x_4 = 2 - - - - - - - - -(2) \\ -x_4 = 3. - - - - - - - - - - - - - -(3) \end{cases}$$

(*a*) Find the coefficient matrix and the augmented matrix of the system.
(*b*) Find the basic variables and free variables of the system.
(*c*) Solve the system.

(d) If the system is consistent, then express its solution as a linear combination.

(e) Determine if the vector $\overrightarrow{b} = \begin{pmatrix} 1 \\ 2 \\ 3 \end{pmatrix}$ is a linear combination of the column vectors of the coefficient matrix of the system.

Solution. (a) The coefficient matrix and the augmented matrix are

$$A = \begin{pmatrix} 1 & -1 & -3 & 1 \\ 0 & -1 & -1 & 1 \\ 0 & 0 & 0 & -1 \end{pmatrix} \quad \text{and } (A|\overrightarrow{b}) = \left(\begin{array}{cccc|c} 1 & -1 & -3 & 1 & 1 \\ 0 & -1 & -1 & 1 & 2 \\ 0 & 0 & 0 & -1 & 3 \end{array}\right).$$

(b) The coefficient matrix is a row echelon matrix. The variables corresponding to the leading entries are x_1, x_2 and x_4, respectively and x_3 is a free variable.

(c) Since x_3 is a free variable. Let $x_3 = t$, where $t \in \mathbb{R}$. By (3), we have $x_4 = -3$. By (2), we have

$$x_2 = -2 - x_3 + x_4 = -2 - t + (-3) = -5 - t.$$

By (1), we obtain

$$x_1 = 1 + x_2 + 3x_3 - x_4 = 1 + (-5 - t) + 3t - (-3) = -1 + 2t.$$

Hence,

$$\begin{pmatrix} x_1 \\ x_2 \\ x_3 \\ x_4 \end{pmatrix} = \begin{pmatrix} -1 + 2t \\ -5 - t \\ t \\ -3 \end{pmatrix}. \tag{4.13}$$

(d) We can write the solution (4.13) as a linear combination:

$$\begin{pmatrix} x_1 \\ x_2 \\ x_3 \\ x_4 \end{pmatrix} = \begin{pmatrix} -1 + 2t \\ -5 - t \\ 0 + t \\ -3 + 0t \end{pmatrix} = \begin{pmatrix} -1 \\ -5 \\ 0 \\ -3 \end{pmatrix} + \begin{pmatrix} 2t \\ -t \\ t \\ 0t \end{pmatrix}$$
$$= \begin{pmatrix} -1 \\ -5 \\ 0 \\ -3 \end{pmatrix} + t \begin{pmatrix} 2 \\ -1 \\ 1 \\ 0 \end{pmatrix}.$$

(e) Since the system has infinite many solutions, so $\overrightarrow{b} = \begin{pmatrix} 1 \\ 2 \\ 3 \end{pmatrix}$ is a linear combination of the column vectors of the coefficient matrix A. □

Remark 4.4.4. The above example shows that a system has infinite many solutions if the following conditions hold.

(1) The leading entries are less than the numbers of variables.

(2) The coefficient matrix and the augmented matrix have same numbers of nonzero rows,

Example 4.4.5. Consider the following system

$$\begin{cases} x_1 - x_2 - 3x_3 + x_4 - x_5 = 2 ------(1) \\ -x_3 + x_4 = 3 -------------(2) \\ -x_4 + x_5 = 1. -------------(3) \end{cases}$$

(*a*) Find the coefficient matrix and the augmented matrix of the system.

(*b*) Find the basic variables and free variables of the system.

(*c*) Solve the system.

(*d*) Express the solutions as a linear combination.

(*e*) Determine if the vector $\overrightarrow{b} = \begin{pmatrix} 2 \\ 3 \\ 1 \end{pmatrix}$ is a linear combination of the column vectors of the coefficient matrix of the system.

Solution. (*a*) The coefficient matrix is

$$A = \begin{pmatrix} 1 & -1 & -3 & 1 & -1 \\ 0 & 0 & -1 & 1 & 0 \\ 0 & 0 & 0 & -1 & 1 \end{pmatrix}$$

and the augmented matrix is

$$(A|\overrightarrow{b}) = \left(\begin{array}{ccccc|c} 1 & -1 & -3 & 1 & -1 & 2 \\ 0 & 0 & -1 & 1 & 0 & 3 \\ 0 & 0 & 0 & -1 & 1 & 1 \end{array}\right).$$

(*b*) The coefficient matrix A is a row echelon matrix. The variables corresponding to the leading entries are x_1, x_3 and x_4. x_2 and x_5 are free variables.

(*c*) Since x_2 and x_5 are free variables. Let $x_2 = s$ and $x_5 = t$, where $s, t \in \mathbb{R}$. By (3), we have $x_4 = -1 + x_5 = -1 + t$. By (2), we have

$$x_3 = -3 + x_4 = -3 + (-1 + t) = -4 + t.$$

By (1), we obtain

$$x_1 = 2 + x_2 + 3x_3 - x_4 + x_5 = 2 + s + 3(-4 + t) - (-1 + t) + t = -9 + s + 3t.$$

Hence, $\begin{pmatrix} x_1 \\ x_2 \\ x_3 \\ x_4 \\ x_5 \end{pmatrix} = \begin{pmatrix} -9+s+3t \\ s \\ -4+t \\ -1+t \\ t \end{pmatrix}$.

(d) We write the above solution as a linear combination:

$$\begin{pmatrix} x_1 \\ x_2 \\ x_3 \\ x_4 \\ x_5 \end{pmatrix} = \begin{pmatrix} -9+s+3t \\ 0+s+0t \\ -4+0s+t \\ -1+0s+t \\ 0+0s+t \end{pmatrix} = \begin{pmatrix} -9 \\ 0 \\ -4 \\ -1 \\ 0 \end{pmatrix} + \begin{pmatrix} s \\ s \\ 0 \\ 0 \\ 0 \end{pmatrix} + \begin{pmatrix} 3t \\ 0t \\ t \\ t \\ t \end{pmatrix}$$

$$= \begin{pmatrix} -9 \\ 0 \\ -4 \\ -1 \\ 0 \end{pmatrix} + s\begin{pmatrix} 1 \\ 1 \\ 0 \\ 0 \\ 0 \end{pmatrix} + t\begin{pmatrix} 3 \\ 0 \\ 1 \\ 1 \\ 1 \end{pmatrix}.$$

(e) Since the system has infinite many solutions, so $\overrightarrow{b} = \begin{pmatrix} 2 \\ 3 \\ 1 \end{pmatrix}$ is a linear combination of the column vectors of the coefficient matrix A. □

Example 4.4.6. Consider the following system

$$\begin{cases} x_1 - x_2 - 3x_3 = 1 ------(1) \\ -x_2 + x_3 = 3 --------(2) \\ 0x_3 = 1. ----------(3) \end{cases}$$

(a) Find the coefficient matrix and the augmented matrix of the system.
(b) Find the basic variables and free variables of the system.
(c) Solve the system.

Solution. (a) The coefficient matrix and the augmented matrix are

$$A = \begin{pmatrix} 1 & -1 & -3 \\ 0 & -1 & 1 \\ 0 & 0 & 0 \end{pmatrix} \quad \text{and } (A|\overrightarrow{b}) = \left(\begin{array}{ccc|c} 1 & -1 & -3 & 1 \\ 0 & -1 & 1 & 3 \\ 0 & 0 & 0 & 1 \end{array}\right).$$

(b) The coefficient matrix A is a row echelon matrix. The variables corresponding to the leading entries are x_1 and x_2, respectively and x_3 is a free variable.

(c) Since the last equation (3) implies $0 = 1$, a contradiction. Hence, the system has no solutions. □

Remark 4.4.7. Example 4.4.6 shows that a system has no solutions if the following conditions hold.

(i) The leading entries are less than the numbers of variables.

(ii) The number of nonzero rows of the coefficient matrix is less than the number of nonzero rows of the augmented matrix.

Exercises

1. Consider the system of the following linear equations

$$\begin{cases} x_1 + x_2 - 3x_3 = 2 \\ -x_2 - 6x_3 = 4 \\ -x_3 = 1. \end{cases}$$

(a) Find the coefficient matrix and the augmented matrix of the system.

(b) Find the basic variables and free variables of the system.

(c) Solve the system by using back-substitution.

(d) Determine if the vector $\overrightarrow{b} = \begin{pmatrix} 2 \\ 4 \\ 1 \end{pmatrix}$ is a linear combination of the column vectors of the coefficient matrix of the system.

2. Consider the following system

$$\begin{cases} x_1 - 2x_2 + 3x_3 + x_4 = -1 \\ x_2 - x_3 + x_4 = 4 \\ -x_4 = 2. \end{cases}$$

(a) Find the coefficient matrix and the augmented matrix of the system.

(b) Find the basic variables and free variables of the system.

(c) Solve the system by using back-substitution.

(d) If the system is consistent, then express its solution as a linear combination.

(e) Determine if the vector $\overrightarrow{b} = \begin{pmatrix} -1 \\ 4 \\ 2 \end{pmatrix}$ is a linear combination of the column vectors of the coefficient matrix of the system.

3. Consider the following system

$$\begin{cases} -x_1 - 2x_2 - 2x_3 + x_4 - 3x_5 = 4 \\ -2x_3 + 3x_4 = 1 \\ -4x_4 + 2x_5 = 5. \end{cases}$$

(*a*) Find the coefficient matrix and the augmented matrix of the system.

(*b*) Find the basic variables and free variables of the system.

(*c*) Solve the system.

(*d*) Express the solutions as a linear combination.

(*e*) Determine if the vector $\overrightarrow{b} = \begin{pmatrix} 4 \\ 1 \\ 5 \end{pmatrix}$ is a linear combination of the column vectors of the coefficient matrix of the system.

4. Consider the following system

$$\begin{cases} -3x_1 - 2x_2 - 2x_3 = 2 \\ x_2 - x_3 = 2 \\ 0x_3 = 4. \end{cases}$$

(*a*) Find the coefficient matrix and the augmented matrix of the system.

(*b*) Find the basic variables and free variables of the system.

(*c*) Solve the system.

4.5 Gaussian Elimination

Consider the system (4.5), that is,

$$\begin{cases} a_{11}x_1 + a_{12}x_2 + \cdots + a_{1n}x_n = b_1 \\ a_{21}x_1 + a_{22}x_2 + \cdots + a_{2n}x_n = b_2 \\ \vdots \\ a_{m1}x_1 + a_{m2}x_2 + \cdots + a_{mn}x_n = b_m. \end{cases} \tag{4.14}$$

Let A and $(A|\overrightarrow{b})$ be the coefficient matrix given in (4.6) and the augmented matrix defined in (4.7) of the above system. To solve the system, we follow the following steps:

Step 1: Use row operations to change $(A|\overrightarrow{b})$ to $(B|\overrightarrow{c})$, where B is a row echelon matrix of A.

Step 2. Use the back-substitution to solve the new system of linear equations corresponding to the augmented matrix $(B|\overrightarrow{c})$.

The solutions to the new system are the solutions of (4.14). This method is called the Gaussian elimination.

Example 4.5.1. Solve the system

$$\begin{cases} 2x_1 + 4x_2 + 6x_3 = 18 \\ 4x_1 + 5x_2 + 6x_3 = 24 \\ 3x_1 + x_2 - 2x_3 = 4. \end{cases}$$

Solution.

$$\begin{aligned} (A|\overrightarrow{b}) &= \left(\begin{array}{ccc|c} 2 & 4 & 6 & 18 \\ 4 & 5 & 6 & 24 \\ 3 & 1 & -2 & 4 \end{array}\right) \xrightarrow{R_1(\frac{1}{2})} \left(\begin{array}{ccc|c} 1 & 2 & 3 & 9 \\ 4 & 5 & 6 & 24 \\ 3 & 1 & -2 & 4 \end{array}\right) \\ & \xrightarrow[R_1(-3)+R_3]{R_1(-4)+R_2} \left(\begin{array}{ccc|c} 1 & 2 & 3 & 9 \\ 0 & -3 & -6 & -12 \\ 0 & -5 & -11 & -23 \end{array}\right) \xrightarrow{R_2(-\frac{1}{3})} \\ & \left(\begin{array}{ccc|c} 1 & 2 & 3 & 9 \\ 0 & 1 & 2 & 4 \\ 0 & -5 & -11 & -23 \end{array}\right) \xrightarrow{R_2(5)+R_3} \left(\begin{array}{ccc|c} 1 & 2 & 3 & 9 \\ 0 & 1 & 2 & 4 \\ 0 & 0 & -1 & -3 \end{array}\right) \\ &= (B|\overrightarrow{c}). \end{aligned}$$

The system of linear equations corresponding to $(B|\overrightarrow{c})$ is

$$\begin{cases} x_1 + 2x_2 + 3x_3 = 9 - - - - - - - - (1) \\ x_2 + 2x_3 = 4 - - - - - - - - - - - -(2) \\ -x_3 = -3. - - - - - - - - - - - - - (3) \end{cases}$$

Now, we use the back-substitution to solve the above system. By (3), we have $x_3 = 3$. By (2), we have $x_2 = 4 - 2x_3 = 4 - 2(3) = -2$. By (1), we obtain

$$x_1 = 9 - 2x_2 - 3x_3 = 9 - 2(-2) - 3(3) = 4.$$

Hence $(x_1, x_2, x_3) = (4, -2, 3)$ is a solution of the system. □

Example 4.5.2. Solve the system

$$\begin{cases} 2x_1 + 2x_2 - 2x_3 = 4 \\ 3x_1 + 5x_2 + x_3 = -8 \\ -4x_1 - 7x_2 - 2x_3 = 13 \end{cases}$$

and express the solutions as a linear combination.

Solution.

$$(A|\overrightarrow{b}) = \left(\begin{array}{ccc|c} 2 & 2 & -2 & 4 \\ 3 & 5 & 1 & -8 \\ -4 & -7 & -2 & 13 \end{array}\right) \xrightarrow{R_1(\frac{1}{2})} \left(\begin{array}{ccc|c} 1 & 1 & -1 & 2 \\ 3 & 5 & 1 & -8 \\ -4 & -7 & -2 & 13 \end{array}\right)$$

$$\xrightarrow[R_1(4)+R_3]{R_1(-3)+R_2} \left(\begin{array}{ccc|c} 1 & 1 & -1 & 2 \\ 0 & 2 & 4 & -14 \\ 0 & -3 & -6 & 21 \end{array}\right) \xrightarrow{R_2(\frac{1}{2})}$$

$$\left(\begin{array}{ccc|c} 1 & 1 & -1 & 2 \\ 0 & 1 & 2 & -7 \\ 0 & -3 & -6 & 21 \end{array}\right) \xrightarrow{R_2(3)+R_3} \left(\begin{array}{ccc|c} 1 & 1 & -1 & 2 \\ 0 & 1 & 2 & -7 \\ 0 & 0 & 0 & 0 \end{array}\right)$$

$$= (B|\overrightarrow{c}).$$

The system of linear equations corresponding to $(B|\overrightarrow{c})$ is

$$\begin{cases} x_1 + x_2 - x_3 = 2 - - - - - - - - (1) \\ x_2 + 2x_3 = -7. - - - - - - - - -(2) \end{cases}$$

Now, we use the back-substitution to solve the above system. x_1 and x_2 are the basic variables and x_3 is a free variable. Let $x_3 = t$, where $t \in \mathbb{R}$. By (2), we get $x_2 = -7 - 2x_3 = -7 - 2t$ and by (1),

$$x_1 = 2 - x_2 + x_3 = 2 - (-7 - 2t) + t = 9 + 3t.$$

The solution is

$$\begin{pmatrix} x_1 \\ x_2 \\ x_3 \end{pmatrix} = \begin{pmatrix} 9+3t \\ -7-2t \\ t \end{pmatrix}.$$

We write the above solution as a linear combination:

$$\begin{pmatrix} x_1 \\ x_2 \\ x_3 \end{pmatrix} = \begin{pmatrix} 9+3t \\ -7-2t \\ t \end{pmatrix} = \begin{pmatrix} 9 \\ -7 \\ 0 \end{pmatrix} + t \begin{pmatrix} 3 \\ -2 \\ 1 \end{pmatrix}.$$

□

Example 4.5.3. Solve the following system

$$\begin{cases} x_1 + 2x_2 + 3x_3 = 4 \\ 4x_1 + 7x_2 + 6x_3 = 17 \\ 2x_1 + 5x_2 + 12x_3 = 10. \end{cases}$$

Solution.

$$(A|\overrightarrow{b}) = \left(\begin{array}{ccc|c} 1 & 2 & 3 & 4 \\ 4 & 7 & 6 & 17 \\ 2 & 5 & 12 & 10 \end{array}\right) \xrightarrow[R_1(-2)+R_3]{R_1(-4)+R_2} \left(\begin{array}{ccc|c} 1 & 2 & 3 & 4 \\ 0 & -1 & -6 & 1 \\ 0 & 1 & 6 & 2 \end{array}\right)$$

$$\xrightarrow{R_2(1)+R_3} \left(\begin{array}{ccc|c} 1 & 2 & 3 & 4 \\ 0 & -1 & -6 & 1 \\ 0 & 0 & 0 & 3 \end{array}\right) = (B|\overrightarrow{c}).$$

The system of linear equations corresponding to $(B|\overrightarrow{c})$ is

$$\begin{cases} x_1 + 2x_2 + 3x_3 = 4 - - - - - - -(1) \\ -x_2 - 6x_3 = 1 - - - - - - - - -(2) \\ 0x_1 + 0x_2 + 0x_3 = 3. - - - - - - (3) \end{cases}$$

The last equation (3) reads $0 = 3$, this is impossible. Thus the orginal system has no solutions. □

The above examples can be used to verify the following theorem:

Theorem 4.5.4. *Let A be the $m \times n$ coefficient matrix of (4.14) and let $(A|\overrightarrow{b})$ be the augmented matrix of (4.14). Assume that B and $(B|\overrightarrow{c})$ are row echelon matrices of A and $(A|\overrightarrow{b})$, respectively, obtained by using the row operations. Then the following assertions hold.*

(1) *If the number of the leading entries of B is equal to the number of variables, then the system (4.14) has a unique solution.*

(2) *If the number of the leading entries of B is less than the number of variables and the number of nonzero rows of B is equal to the number of nonzero rows of $(B|\overrightarrow{c})$, then the system (4.14) has infinite many solutions.*

(3) *If the number of the leading entries of B is less than the number of variables and the number of nonzero rows of B is less than the number of nonzero rows of $(B|\overrightarrow{c})$, then the system (4.14) has no solutions.*

Exercises

1. Use Gaussian elimination to solve the following systems and express the solutions as linear combinations if they have infinitely many solutions.

a) $\begin{cases} x_1 - 2x_2 + 3x_3 = 6 \\ 2x_1 + x_2 + 4x_3 = 5 \\ -3x_1 + x_2 - 2x_3 = 3. \end{cases}$ b) $\begin{cases} x_1 + x_2 - 2x_3 = 3 \\ 2x_1 + 3x_2 - x_3 = -4 \\ -2x_1 - 3x_2 - x_3 = 4 \end{cases}$

c) $\begin{cases} x_1 + 2x_2 + 3x_3 = 4 \\ 4x_1 + 7x_2 + 6x_3 = 17 \\ 2x_1 + 5x_2 + 12x_3 = 7. \end{cases}$

4.6 Basic variables, free variables and solutions

Let A be an $m \times n$ matrix and B its row echelon matrix. By Definition 2.13.1, the rank of A is equal to the number of nonzero rows of B. Equivalently, the rank of A is equal to the number of the basic variables of B. Hence, by Theorem 4.5.4, we have the following result which provides the relation between the rank of A and solutions of a system of linear equations.

Theorem 4.6.1. *Let A and $(A|\overrightarrow{b})$ be the coefficient matrix and the augmented matrix of the system (4.14), respectively. Then the following assertions hold.*

(1) *If $r(A) = n$, then the system (4.14) has a unique solution.*

(2) *If $r(A) < n$ and $r(A) = r(A|\overrightarrow{b})$, then the system (4.14) has infinite many solutions.*

(3) *If $r(A) < n$ and $r(A) < r(A|\overrightarrow{b})$, then the system (4.14) has no solutions.*

Theorem 4.6.2. *For a system of linear equations of (4.14), one of the following assertions must occur.*

(1) *(4.14) has a unique solution.*
(2) *(4.14) has infinite many solutions.*
(3) *(4.14) has no solutions.*

By Theorem 4.6.2, we obtain the following result on homogeneous systems.

Theorem 4.6.3. *Let A be the coefficient matrix of the homogeneous system (4.8). Then the following assertions hold.*

(1) *If $r(A) = n$, then the homogeneous system (4.8) has a unique solution.*

(2) *If $r(A) < n$, then the homogeneous system (4.8) has infinite many solutions.*

Proof. Let $(A|\overrightarrow{0})$ be the augmented matrix of (4.8). Then $r(A) = (A|\overrightarrow{0})$. The result follows from Theorem 4.6.2. □

By Theorems 4.6.3 and 2.13.5, we have

Corollary 4.6.4. *One of the following assertions must occur.*

(1) *The homogeneous system (4.8) has a unique solution. In this case, the unique solution must be the zero solution.*

(2) *The homogeneous system (4.8) has infinite many solutions.*

Corollary 4.6.5. *In the homogeneous system (4.8), if $n > m$, then the homogeneous system (4.8) has infinitely many solutions.*

Proof. By Theorem 2.13.5, we have $r(A) \leq \min\{n, m\}$. Since $m < n$, we have

$$r(A) \leq \min\{n, m\} = m < n.$$

The result follows from Theorem 4.6.3 (2). $\square$

Remark 4.6.6. In (4.8), if $n \leq m$, it is possible that the system (4.8) has either only the zero solution or infinitely many solutions.

According to Theorem 4.6.1 and Definition 4.3.7, a system of linear equations is consistent if it has a unique solution or infinite many solutions.

Exercises

1. Use Theorem 4.6.1 to determine whether the following systems have a unique solution, infinitely many solutions or no solutions.

a) $\begin{cases} x_1 - 2x_2 + 3x_3 = 6 \\ 2x_1 + x_2 + 4x_3 = 5 \\ -3x_1 + x_2 - 2x_3 = 3. \end{cases}$ $b)$ $\begin{cases} x_1 + x_2 - 2x_3 = 3 \\ 2x_1 + 3x_2 - x_3 = -4 \\ -2x_1 - 3x_2 - x_3 = 4 \end{cases}$

c) $\begin{cases} x_1 + 2x_2 + 3x_3 = 4 \\ 4x_1 + 7x_2 + 6x_3 = 17 \\ 2x_1 + 5x_2 + 12x_3 = 7. \end{cases}$ $d)$ $\begin{cases} x_1 - 2x_2 - x_3 + 2x_4 = 0 \\ -x_2 - 2x_3 + x_4 = 0. \end{cases}$

e) $\begin{cases} x_1 - x_2 + 3x_3 = 2 \\ -2x_1 + 2x_2 - 6x_3 = 5 \\ 2x_1 - 2x_2 + 6x_3 = 4. \end{cases}$

4.7 Consistent linear systems

Consider the system (4.14), that is,

$$\begin{cases} a_{11}x_1 + a_{12}x_2 + \cdots + a_{1n}x_n = b_1 \\ a_{21}x_1 + a_{22}x_2 + \cdots + a_{2n}x_n = b_2 \\ \vdots \\ a_{m1}x_1 + a_{m2}x_2 + \cdots + a_{mn}x_n = b_m. \end{cases} \tag{4.15}$$

Let A and $(A|\overrightarrow{b})$ be the coefficient matrix given in (4.6) and the augmented matrix defined in (4.7) of the system (4.15), respectively.

We want to determine for what vectors $\overrightarrow{b} = \begin{pmatrix} b_1 \\ b_2 \\ \vdots \\ b_m \end{pmatrix}$, the system (4.15) is consistent or inconsistent, that is, find the relations on $b_1, b_2, \cdots, b_m$ such that (4.15) has a unique solution, or has infinite many solutions or has no solutions.

The method is to use row operations to change $(A|\overrightarrow{b})$, where we treat $b_1, b_2, \cdots, b_m$ as constants, to a row echelon matrix $(B|\overrightarrow{c})$ and then to use Theorem 4.6.1 to find the conditions on $b_1, b_2, \cdots, b_m$.

Example 4.7.1. Consider the following system of linear equations

$$\begin{cases} x_1 + 2x_2 + 3x_3 = b_1 \\ 2x_1 + 5x_2 + 4x_3 = b_2 \\ x_1 + 5x_3 = b_3. \end{cases}$$

Find conditions on b_1, b_2, b_3 such that the system is consistent.

Solution.

$$(A|\overrightarrow{b}) = \left(\begin{array}{ccc|c} 1 & 2 & 3 & b_1 \\ 2 & 5 & 4 & b_2 \\ 1 & 0 & 5 & b_3 \end{array}\right) \xrightarrow[R_1(-1)+R_3]{R_1(-2)+R_2} \left(\begin{array}{ccc|c} 1 & 2 & 3 & b_1 \\ 0 & 1 & -2 & b_2 - 2b_1 \\ 0 & -2 & 5 & b_3 - b_1 \end{array}\right)$$

$$\xrightarrow{R_2(2)+R_3} \left(\begin{array}{ccc|c} 1 & 2 & 3 & b_1 \\ 0 & 1 & -2 & b_2 - 2b_1 \\ 0 & 0 & 1 & (b_3 - b_1) + 2(b_2 - 2b_1) \end{array}\right)$$

$$= \left(\begin{array}{ccc|c} 1 & 2 & 3 & b_1 \\ 0 & 1 & -2 & b_2 - 2b_1 \\ 0 & 0 & 1 & b_3 + 2b_2 - 3b_1 \end{array}\right) = (B|\overrightarrow{c}).$$

From the row echelon matrix $(B|\overrightarrow{c})$, we see that the number of nonzero rows of B is 3, so $r(A) = 3$. Since A is a 3×3 matrix, so $n = 3$. Hence, $r(A) = n = 3$. It follows from Theorem 4.6.1 that the system has a unique solution for any $b_1, b_2, b_3 \in \mathbb{R}$. Hence, the system is consistent for $b_1, b_2, b_3 \in \mathbb{R}$. □

From the above example, we see that $r(A) = r(A|\overrightarrow{b})$ and the system has a unique solution for any b_1, b_2, b_3. In fact, one can prove that it is true for any system of linear equations.

Theorem 4.7.2. *For a system of linear equations of* (4.15), *if* $r(A) = r(A|\overrightarrow{b})$ *for any* b_1, b_2, b_3, *then the system has a unique solution.*

Example 4.7.3. Find b_1, b_2, b_3 such that the following system is consistent.

$$\begin{cases} x_1 - 2x_2 + x_3 = b_1 \\ -4x_1 + 5x_2 + 2x_3 = b_2 \\ 4x_1 - 5x_2 + 3x_3 = b_3 \end{cases}$$

Solution.

$$(A|\overrightarrow{b}) = \left(\begin{array}{ccc|c} 1 & -2 & 1 & b_1 \\ -4 & 5 & 2 & b_2 \\ 4 & -5 & 3 & b_3 \end{array}\right) \xrightarrow[R_1(-4)+R_3]{R_1(4)+R_2}$$

$$\left(\begin{array}{ccc|c} 1 & -2 & -1 & b_1 \\ 0 & -3 & 6 & b_2+4b_1 \\ 0 & 3 & -1 & b_3-4b_1 \end{array}\right) \xrightarrow{R_2(1)+R_3}$$

$$\left(\begin{array}{ccc|c} 1 & -2 & -1 & b_1 \\ 0 & -3 & 6 & b_2+4b_1 \\ 0 & 0 & 5 & b_3+b_2 \end{array}\right) = (B|\overrightarrow{c}).$$

Hence, $r(A) = r(B|\overrightarrow{c})$ for any b_1, b_2, b_3. Hence the system is consistent. □

From the above examples, we see that $r(A) = r(A|\overrightarrow{b})$ is independent of b_1, b_2, b_3, so the system has a unique solution. The following examples show that in some cases, $r(A|\overrightarrow{b})$ is dependent of b_1, b_2, b_3, which implies that the system has infinite many solutions or has no solutions.

Example 4.7.4. Consider the following system

$$\begin{cases} 6x_1 - 4x_2 = b_1 \\ 3x_1 - 2x_2 = b_2. \end{cases}$$

(1) Find b_1 and b_2 such that the system is consistent.
(2) Find b_1 and b_2 such that the system is inconsistent.
(3) Determine whether the system with $\overrightarrow{b} = \begin{pmatrix} 4 \\ 1 \end{pmatrix}$ is inconsistent.

Solution.

$$(A|\overrightarrow{b}) = \left(\begin{array}{cc|c} 6 & -4 & b_1 \\ 3 & -2 & b_2 \end{array}\right) \xrightarrow{R_1(\frac{1}{6})} \left(\begin{array}{cc|c} 1 & -\frac{2}{3} & \frac{b_1}{6} \\ 3 & -2 & b_2 \end{array}\right)$$

$$\xrightarrow{R_1(-3)+R_2} \left(\begin{array}{cc|c} 1 & -\frac{2}{3} & \frac{b_1}{6} \\ 0 & 0 & b_2 - \frac{b_1}{2} \end{array}\right).$$

(1) If $b_2 - \frac{b_1}{2} = 0$, this is, $b_1 = 2b_2$, then $r(A) = (A|\overrightarrow{b}) = 1 < 2$. It follows from Theorem 4.6.1 (2), the system has infinite many solutions and is consistent.

(2) If $b_2 - \frac{b_1}{2} \neq 0$, this is, $b_1 \neq 2b_2$, then $r(A) = 1$ and $(A|\overrightarrow{b}) = 2$. Hence, $r(A) < (A|\overrightarrow{b})$. It follows from Theorem 4.6.1 (3), the system has no solutions and is inconsistent.

(3) By $\overrightarrow{b} = \begin{pmatrix} 4 \\ 1 \end{pmatrix}$, we have $b_1 = 4$ and $b_2 = 1$. Since $b_1 \neq 2b_2$, it follows from the result (2) that the system is inconsistent. □

Example 4.7.5. Consider the system of linear equations

$$\begin{cases} x_1 + x_2 + 2x_3 = b_1 \\ x_1 + x_3 = b_2 \\ 3x_1 + 4x_2 + 7x_3 = b_3. \end{cases}$$

(1) Find conditions on b_1, b_2, b_3 such that the system is consistent.
(2) Find conditions on b_1, b_2, b_3 such that the system is inconsistent.
(3) Find a vector $\overrightarrow{b} = \begin{pmatrix} b_1 \\ b_2 \\ b_3 \end{pmatrix}$ such that the system with the vector $\overrightarrow{b}$ is inconsistent.

Solution.

$$(A|\overrightarrow{b}) = \left(\begin{array}{ccc|c} 1 & 1 & 2 & b_1 \\ 1 & 0 & 1 & b_2 \\ 3 & 4 & 7 & b_3 \end{array}\right) \xrightarrow[R_1(-3)+R_3]{R_1(-1)+R_2} \left(\begin{array}{ccc|c} 1 & 1 & 2 & b_1 \\ 0 & -1 & -1 & b_2 - b_1 \\ 0 & 1 & 1 & b_3 - 3b_1 \end{array}\right)$$

$$\xrightarrow{R_2(1)+R_3} \left(\begin{array}{ccc|c} 1 & 1 & 2 & b_1 \\ 0 & 1 & 1 & b_1 - b_2 \\ 0 & 0 & 0 & -4b_1 + b_2 + b_3 \end{array}\right) = (B|\overrightarrow{c}).$$

(1) By Theorem 4.6.1, if $-4b_1 + b_2 + b_3 = 0$, that is, $b_3 = 4b_1 - b_2$, then the system is consistent
(2) Similarly, if $b_3 \neq 4b_1 - b_2$, then the system is inconsistent.
(3) Let $b_1 = b_2 = 0$ and $b_3 = 1$. Then $b_3 \neq 4b_1 - b_2$. Hence, the system with $\overrightarrow{b} = \begin{pmatrix} 0 \\ 0 \\ 1 \end{pmatrix}$ is inconsistent. □

Exercises

1. Find conditions on b_1, b_2, b_3 such that each of the systems is consistent.

a) $\begin{cases} x_1 - x_2 + 2x_3 = b_1 \\ 2x_1 + 4x_2 - 4x_3 = b_2 \\ -3x_1 + 4x_3 = b_3. \end{cases}$ b) $\begin{cases} x_1 - 2x_2 + 2x_3 = b_1 \\ -3x_1 + 3x_2 + x_3 = b_2 \\ 4x_1 - 6x_2 + 3x_3 = b_3. \end{cases}$

2. Consider the following system

$$\begin{cases} 2x_1 - 4x_2 = b_1 \\ x_1 - 2x_2 = b_2. \end{cases}$$

(1) Find b_1 and b_2 such that the system is consistent.

(2) Find b_1 and b_2 such that the system is inconsistent.

(3) If $b_1 = 4$ and $b_2 = 1$, determine whether the system is inconsistent.

3. Consider the system of linear equations

$$\begin{cases} x_1 - x_2 + x_3 = b_1 \\ -2x_1 - x_3 = b_2 \\ 2x_1 - 10x_2 + 6x_3 = b_3. \end{cases}$$

(1) Find conditions on b_1, b_2, b_3 such that the system is consistent.

(2) Find conditions on b_1, b_2, b_3 such that the system is inconsistent.

(3) Find a vector (b_1, b_2, b_3) such that the system is inconsistent.

4.8 Gauss-Jordan Elimination

Consider the system (4.15), that is,

$$\begin{cases} a_{11}x_1 + a_{12}x_2 + \cdots + a_{1n}x_n = b_1 \\ a_{21}x_1 + a_{22}x_2 + \cdots + a_{2n}x_n = b_2 \\ \quad\vdots \\ a_{m1}x_1 + a_{m2}x_2 + \cdots + a_{mn}x_n = b_m. \end{cases} \tag{4.16}$$

Let A and $(A|\overrightarrow{b})$ be the coefficient matrix given in (4.6) and the augmented matrix defined in (4.7) of the system (4.16), respectively.

We have studied the Gaussian Elimination, where the row operations are used to change $(A|\overrightarrow{b})$ to its row echelon matrix $(B|\overrightarrow{c})$ and then the back substitution is used to find the solutions of the new system corresponding to the augmented matrix $(B|\overrightarrow{c})$, where B is the row echelon matrix of A.

Sometimes, we do not need to use the back-substitution, but change the matrix A into a reduced row matrix D directly, that is, we can use row operations to change $(A|\overrightarrow{b})$ to its reduced row echelon matrix $(D|\overrightarrow{d})$, where D is the reduced row echelon matrix of A. Then we solve the new system corresponding to the augmented matrix $(D|\overrightarrow{d})$. The solutions of the new system corresponding to $(D|\overrightarrow{d})$ are solutions of the original system (4.16).

The method mentioned above is called the Gauss-Jordan elimination. We can simply express the Gauss-Jordan elimination as

$$(A|\overrightarrow{b}) \to (B|\overrightarrow{c}) \to (D|\overrightarrow{d}),$$

where B is a row echelon matrix of A and D is the reduced echelon matrix of A.

Example 4.8.1. Solve the following system using Gauss-Jordan elimination.

$$\begin{cases} 2x_1 + 4x_2 + 6x_3 = 18 \\ 4x_1 + 5x_2 + 6x_3 = 24 \\ 3x_1 + x_2 - 2x_3 = 4. \end{cases}$$

Solution. Step 1. Change $(A|\overrightarrow{b})$ to a row echelon matrix $(B|\overrightarrow{c})$. By the solution of Example 4.5.1, we have the row echelon matrix

$$(B|\overrightarrow{c}) = \left(\begin{array}{ccc|c} 1 & 2 & 3 & 9 \\ 0 & 1 & 2 & 4 \\ 0 & 0 & -1 & -3 \end{array}\right).$$

Step 2. Use the row operations to change $(B|\overrightarrow{c})$ to $(D|\overrightarrow{d})$, where D is a reduced row echelon matrix of A.

$$(B|\overrightarrow{c}) = \left(\begin{array}{ccc|c} 1 & 2 & 3 & 9 \\ 0 & 1 & 2 & 4 \\ 0 & 0 & -1 & -3 \end{array}\right) \xrightarrow{R_3(-1)} \left(\begin{array}{ccc|c} 1 & 2 & 3 & 9 \\ 0 & 1 & 2 & 4 \\ 0 & 0 & 1 & 3 \end{array}\right)$$

$$\xrightarrow[R_2(-2)+R_2]{R_3(-3)+R_1} \left(\begin{array}{ccc|c} 1 & 2 & 0 & 0 \\ 0 & 1 & 0 & -2 \\ 0 & 0 & 1 & 3 \end{array}\right) \xrightarrow{R_2(-2)+R_1} \left(\begin{array}{ccc|c} 1 & 0 & 0 & 4 \\ 0 & 1 & 0 & -2 \\ 0 & 0 & 1 & 3 \end{array}\right)$$

$$= (D|\overrightarrow{d}).$$

The system of linear equations corresponding to $(D|\overrightarrow{d})$ is

$$\begin{cases} x_1 = 4 \\ x_2 = -2 \\ x_3 = 3, \end{cases}$$

which is the solution of the system. □

Example 4.8.2. Solve the following system using Gauss-Jordan elimination.

$$\begin{cases} x_1 + x_2 + 2x_3 = 1 \\ -2x_1 - 4x_2 - 5x_3 = -1 \\ 3x_1 + 6x_2 + 5x_3 = 0. \end{cases}$$

Solution.

$$(A|\overrightarrow{b}) = \left(\begin{array}{ccc|c} 1 & 1 & 2 & 1 \\ -2 & -4 & -5 & -1 \\ 3 & 6 & 5 & 0 \end{array}\right) \xrightarrow[R_1(-3)+R_3]{R_1(2)+R_2} \left(\begin{array}{ccc|c} 1 & 1 & 2 & 1 \\ 0 & -2 & -1 & 1 \\ 0 & 3 & -1 & -3 \end{array}\right)$$

$$\xrightarrow{R_3(1)+R_2} \left(\begin{array}{ccc|c} 1 & 1 & 2 & 1 \\ 0 & 1 & -2 & -2 \\ 0 & 3 & -1 & -3 \end{array}\right) \xrightarrow{R_2(-3)+R_3} \left(\begin{array}{ccc|c} 1 & 1 & 2 & 1 \\ 0 & 1 & -2 & -2 \\ 0 & 0 & 5 & 3 \end{array}\right)$$

$$\xrightarrow{R_3(\frac{1}{5})} \left(\begin{array}{ccc|c} 1 & 1 & 2 & 1 \\ 0 & 1 & -2 & -2 \\ 0 & 0 & 1 & \frac{3}{5} \end{array}\right) \xrightarrow[R_3(2)+R_2]{R_3(-2)+R_1} \left(\begin{array}{ccc|c} 1 & 1 & 0 & -\frac{1}{5} \\ 0 & 1 & 0 & -\frac{4}{5} \\ 0 & 0 & 1 & \frac{3}{5} \end{array}\right)$$

$$\xrightarrow{R_2(-1)+R_1} \left(\begin{array}{ccc|c} 1 & 0 & 0 & \frac{3}{5} \\ 0 & 1 & 0 & -\frac{4}{5} \\ 0 & 0 & 1 & \frac{3}{5} \end{array}\right) = (D|\overrightarrow{d}).$$

The system of linear equations corresponding to $(D|\overrightarrow{d})$ is

$$\begin{cases} x_1 = \frac{3}{5} \\ x_2 = -\frac{4}{5} \\ x_3 = \frac{3}{5}, \end{cases}$$

which is the solution of the system. □

Example 4.8.3. Solve the following system using Gauss-Jordan elimination.

$$\begin{cases} x_1 + x_2 - x_3 = 1 \\ -3x_1 + 5x_2 - x_3 = -1 \\ 3x_1 + 7x_2 - 5x_3 = 4. \end{cases}$$

Solution.

$$(A|\overrightarrow{b}) = \left(\begin{array}{ccc|c} 1 & 1 & -1 & 1 \\ -3 & 5 & -1 & -1 \\ 3 & 7 & -5 & 4 \end{array}\right) \xrightarrow[R_1(-3)+R_3]{R_1(3)+R_2} \left(\begin{array}{ccc|c} 1 & 1 & -1 & 1 \\ 0 & 8 & -4 & 2 \\ 0 & 4 & -2 & 1 \end{array}\right)$$

$$\xrightarrow{R_2(\frac{1}{2})} \left(\begin{array}{ccc|c} 1 & 1 & -1 & 1 \\ 0 & 4 & -2 & 1 \\ 0 & 4 & -2 & 1 \end{array}\right) \xrightarrow{R_2(-1)+R_3} \left(\begin{array}{ccc|c} 1 & 1 & -1 & 1 \\ 0 & 4 & -2 & 1 \\ 0 & 0 & 0 & 0 \end{array}\right)$$

$$\xrightarrow{R_2(\frac{1}{4})} \left(\begin{array}{ccc|c} 1 & 1 & -1 & 1 \\ 0 & 1 & -\frac{1}{2} & \frac{1}{4} \\ 0 & 0 & 0 & 0 \end{array}\right) \xrightarrow{R_2(-1)+R_1} \left(\begin{array}{ccc|c} 1 & 0 & -\frac{1}{2} & \frac{3}{4} \\ 0 & 1 & -\frac{1}{2} & \frac{1}{4} \\ 0 & 0 & 0 & 0 \end{array}\right)$$

$$= (D|\overrightarrow{d}).$$

The system of linear equations corresponding to $(D|\overrightarrow{d})$ is

$$\begin{cases} x_1 - \frac{1}{2}x_3 = \frac{3}{4} \\ x_2 - \frac{1}{2}x_3 = \frac{1}{4}, \end{cases}$$

where x_1, x_2 are basic variables and x_3 is a free variable. Let $x_3 = t$. Then

$$\begin{cases} x_1 = \frac{3}{4} + \frac{3}{2}t \\ x_2 = \frac{1}{4} + \frac{1}{2}t \\ x_3 = t, \end{cases}$$

which is a solution of the system for each $t \in \mathbb{R}$. □

Exercises

1. Solve each of the following systems using Gauss-Jordan elimination.

$$a) \begin{cases} x_1 + 3x_2 - x_3 = 2 \\ 4x_1 - 6x_2 + 6x_3 = 14 \\ -3x_1 - x_2 - 2x_3 = 3. \end{cases} \qquad b) \begin{cases} x_1 - 2x_2 + 2x_3 = 2 \\ -2x_1 + 3x_2 - 4x_3 = -2 \\ -3x_1 + 4x_2 + 6x_3 = 0. \end{cases}$$

$$c) \begin{cases} x_2 - x_3 = 1 \\ -x_1 + 3x_2 - x_3 = -2 \\ 3x_1 + 3x_2 - 2x_3 = 4. \end{cases} \qquad d) \begin{cases} x_1 + 4x_2 + 6x_3 = 0 \\ 2x_1 + 6x_2 + 3x_3 = 0 \\ -3x_1 + x_2 - 2x_3 = 0. \end{cases}$$

$$e) \begin{cases} -x_1 + 2x_2 - x_3 = 0 \\ 3x_1 - 3x_2 - 9x_3 = 0 \\ -2x_1 - x_2 + 12x_3 = 0. \end{cases}$$

4.9 Inverse matrix methods

Consider the system (4.15) with $m = n$, that is,

$$\begin{cases} a_{11}x_1 + a_{12}x_2 + \cdots + a_{1n}x_n = b_1 \\ a_{21}x_1 + a_{22}x_2 + \cdots + a_{2n}x_n = b_2 \\ \vdots \\ a_{n1}x_1 + a_{n2}x_2 + \cdots + a_{nn}x_n = b_n. \end{cases} \tag{4.17}$$

Let A be the coefficient matrix of the system (4.17), that is,

$$A = \begin{pmatrix} a_{11} & a_{12} & \cdots & a_{1n} \\ a_{21} & a_{22} & \cdots & a_{2n} \\ \vdots & \vdots & \ddots & \vdots \\ a_{n1} & a_{n2} & \cdots & a_{nn} \end{pmatrix}. \tag{4.18}$$

Such a matrix can be found in (2.6). By (4.9), the system (4.17) can be written as

$$A\overrightarrow{X} = \overrightarrow{b}, \tag{4.19}$$

where $\overrightarrow{X} = \begin{pmatrix} x_1 \\ x_2 \\ \vdots \\ x_n \end{pmatrix}$ and $\overrightarrow{b} = \begin{pmatrix} b_1 \\ b_2 \\ \vdots \\ b_n \end{pmatrix}$.

Theorem 4.9.1. *If A is invertible, then* (4.5) *has a unique solution*

$$\overrightarrow{X} = A^{-1}\overrightarrow{b}. \tag{4.20}$$

Proof. Since A is invertible, A^{-1} exists and $A^{-1}A = I_n$, where I_n is the identity matrix. Multiplying both sides of (4.19) by A^{-1}, we have

$$A^{-1}(A\overrightarrow{X}) = A^{-1}\overrightarrow{b} \Leftrightarrow (A^{-1}A)\overrightarrow{X} = A^{-1}\overrightarrow{b} \Leftrightarrow I_n\overrightarrow{X} = A^{-1}\overrightarrow{b} \Leftrightarrow \overrightarrow{X} = A^{-1}\overrightarrow{b}.$$

□

Example 4.9.2. Solve the following system using the inverse matrix method

$$\begin{cases} x_1 + 2x_2 + x_3 = 2 \\ 2x_1 + 5x_2 + 2x_3 = 6 \\ x_1 - x_3 = 2. \end{cases}$$

Solution. Let

$$A = \begin{pmatrix} 1 & 2 & 1 \\ 2 & 5 & 2 \\ 1 & 0 & -1 \end{pmatrix} \quad \text{and } \overrightarrow{b} = \begin{pmatrix} 2 \\ 6 \\ 2 \end{pmatrix}.$$

We find A^{-1} by using the method given in section 2.14.

$$(A|I_3) = \left(\begin{array}{ccc|ccc} 1 & 2 & 1 & 1 & 0 & 0 \\ 2 & 5 & 2 & 0 & 1 & 0 \\ 1 & 0 & -1 & 0 & 0 & 1 \end{array}\right) \xrightarrow[R_1(-1)+R_3]{R_1(-2)+R_2}$$

$$\left(\begin{array}{ccc|ccc} 1 & 2 & 1 & 1 & 0 & 0 \\ 0 & 1 & 0 & -2 & 1 & 0 \\ 0 & -2 & -2 & -1 & 0 & 1 \end{array}\right) \xrightarrow{R_2(2)+R_3}$$

$$\left(\begin{array}{ccc|ccc} 1 & 2 & 1 & 1 & 0 & 0 \\ 0 & 1 & 0 & -2 & 1 & 0 \\ 0 & 0 & -2 & -5 & 2 & 1 \end{array}\right) \xrightarrow{R_3(-\frac{1}{2})} \left(\begin{array}{ccc|ccc} 1 & 2 & 1 & 1 & 0 & 0 \\ 0 & 1 & 0 & -2 & 1 & 0 \\ 0 & 0 & 1 & \frac{5}{2} & -1 & -\frac{1}{2} \end{array}\right)$$

$$\xrightarrow{R_3(-1)+R_1} \left(\begin{array}{ccc|ccc} 1 & 2 & 0 & -\frac{3}{2} & 1 & \frac{1}{2} \\ 0 & 1 & 0 & -2 & 1 & 0 \\ 0 & 0 & 1 & \frac{5}{2} & -1 & -\frac{1}{2} \end{array}\right) \xrightarrow{R_2(-2)+R_1}$$

$$\left(\begin{array}{ccc|ccc} 1 & 0 & 0 & \frac{5}{2} & -1 & \frac{1}{2} \\ 0 & 1 & 0 & -2 & 1 & 0 \\ 0 & 0 & 1 & \frac{5}{2} & -1 & -\frac{1}{2} \end{array}\right) = (I_3|A^{-1}).$$

Hence, $A^{-1} = \begin{pmatrix} \frac{5}{2} & -1 & \frac{1}{2} \\ -2 & 1 & 0 \\ \frac{5}{2} & -1 & -\frac{1}{2} \end{pmatrix}$. By (4.20), we have

$$\overrightarrow{X} = \begin{pmatrix} x_1 \\ x_2 \\ x_3 \end{pmatrix} = A^{-1}\overrightarrow{b} = \begin{pmatrix} \frac{5}{2} & -1 & \frac{1}{2} \\ -2 & 1 & 0 \\ \frac{5}{2} & -1 & -\frac{1}{2} \end{pmatrix} \begin{pmatrix} 2 \\ 6 \\ 2 \end{pmatrix} = \begin{pmatrix} 0 \\ 2 \\ -2 \end{pmatrix}.$$

Hence, $x_1 = 0$, $x_2 = 2$, and $x_3 = -2$ is the solution of the system. □

Example 4.9.3. Solve the following system using the inverse matrix method

$$\begin{cases} x_1 + 2x_2 + 3x_3 = 0 \\ 2x_1 + 5x_2 + 3x_3 = 1 \\ x_1 + 8x_3 = -1. \end{cases}$$

Solution. Let

$$A = \begin{pmatrix} 1 & 2 & 3 \\ 2 & 5 & 3 \\ 1 & 0 & 8 \end{pmatrix} \quad \text{and } \overrightarrow{b} = \begin{pmatrix} 5 \\ 3 \\ 17 \end{pmatrix}.$$

We find A^{-1}.

$$(A|I_3) = \left(\begin{array}{ccc|ccc} 1 & 2 & 3 & 1 & 0 & 0 \\ 2 & 5 & 3 & 0 & 1 & 0 \\ 1 & 0 & 8 & 0 & 0 & 1 \end{array}\right) \xrightarrow[R_1(-1)+R_3]{R_1(-2)+R_2}$$

$$\left(\begin{array}{ccc|ccc} 1 & 2 & 3 & 1 & 0 & 0 \\ 0 & 1 & -3 & -2 & 1 & 0 \\ 0 & -2 & 5 & -1 & 0 & 1 \end{array}\right) \xrightarrow{R_2(2)+R_3}$$

$$\left(\begin{array}{ccc|ccc} 1 & 2 & 3 & 1 & 0 & 0 \\ 0 & 1 & -3 & -2 & 1 & 0 \\ 0 & 0 & -1 & -5 & 2 & 1 \end{array}\right) \xrightarrow{R_3(-1)} \left(\begin{array}{ccc|ccc} 1 & 2 & 3 & 1 & 0 & 0 \\ 0 & 1 & -3 & -2 & 1 & 0 \\ 0 & 0 & 1 & 5 & -2 & -1 \end{array}\right)$$

$$\xrightarrow[R_3(3)+R_2]{R_3(-3)+R_1} \left(\begin{array}{ccc|ccc} 1 & 2 & 0 & -14 & 6 & 3 \\ 0 & 1 & 0 & 13 & -5 & -3 \\ 0 & 0 & 1 & 5 & -2 & -1 \end{array}\right) \xrightarrow{R_2(-2)+R_1}$$

$$\left(\begin{array}{ccc|ccc} 1 & 0 & 0 & -40 & 16 & 9 \\ 0 & 1 & 0 & 13 & -5 & -3 \\ 0 & 0 & 1 & 5 & -2 & -1 \end{array}\right) = (I_3|A^{-1}).$$

Hence, $A^{-1} = \begin{pmatrix} -40 & 16 & 9 \\ 13 & -5 & -3 \\ 5 & -2 & -1 \end{pmatrix}$. By (4.20), we obtain

$$\overrightarrow{X} = \begin{pmatrix} x_1 \\ x_2 \\ x_3 \end{pmatrix} = A^{-1}\overrightarrow{b} = \begin{pmatrix} -40 & 16 & 9 \\ 13 & -5 & -3 \\ 5 & -2 & -1 \end{pmatrix} \begin{pmatrix} 0 \\ 1 \\ -1 \end{pmatrix} = \begin{pmatrix} 7 \\ -2 \\ -1 \end{pmatrix}.$$

So, $x_1 = 7$, $x_2 = -2$, and $x_3 = -1$ is the solution of the system. □

Exercises

1. Solve each of the following systems using the inverse matrix method.

   a) $\begin{cases} x_1 + x_2 + x_3 = 2 \\ 2x_1 + 3x_2 + x_3 = 2 \\ x_1 - x_3 = 1 \end{cases}$ *b)* $\begin{cases} x_1 - 2x_2 + 2x_3 = 1 \\ 2x_1 + 3x_2 + 3x_3 = -1 \\ x_1 + 6x_3 = -4 \end{cases}$

   c) $\begin{cases} x_2 + x_3 = 4 \\ x_1 + x_2 + x_4 = 5 \\ 4x_1 + x_2 + x_3 = 4 \\ -x_1 - 2x_2 + 3x_4 = 0 \end{cases}$

4.10 Cramer's rule

Consider the system (4.17) in section 4.9. Let A be the coefficient matrix of the system (4.17) given in (4.18). Assume that $|A| \neq 0$. Then by Theorem 3.8.7, A^{-1} exists. We can use the inverse method to solve the system (4.17) by finding A^{-1}, but we give another method, called the Cramer's rule, to solve the system (4.17) by using the determinants.

For each $i \in \mathcal{I}_n$, we define a matrix A_i by replacing the ith column of A by the vector $\overrightarrow{b}$:

$$A_i = \begin{pmatrix} a_{11} & a_{12} & \cdots & a_{1(i-1)} & b_1 & a_{1(i+1)} & \cdots & a_{1n} \\ a_{21} & a_{22} & \cdots & a_{2(i-1)} & b_2 & a_{2(i+1)} & \cdots & a_{2n} \\ \vdots & \vdots & \ddots & \vdots & \vdots & \vdots & \ddots & \vdots \\ a_{n1} & a_{n2} & \cdots & a_{n(i-1)} & b_n & a_{n(i+1)} & \cdots & a_{nn} \end{pmatrix}. \tag{4.21}$$

Theorem 4.10.1. *Assume that A is the coefficient matrix of the system* (4.17) *given in* (4.18) *and $|A| \neq 0$. Then*

$$x_1 = \frac{|A_1|}{|A|}, \; x_2 = \frac{|A_2|}{|A|}, \cdots, \; x_n = \frac{|A_n|}{|A|} \tag{4.22}$$

is the unique solution of (4.17).

Example 4.10.2. Solve the following system by using Cramer's rule.

$$\begin{cases} 2x - 3y = 7 \\ x + 5y = 1. \end{cases}$$

Solution. Let

$$A = \begin{pmatrix} a_{11} & a_{12} \\ a_{21} & a_{22} \end{pmatrix} = \begin{pmatrix} 2 & -3 \\ 1 & 5 \end{pmatrix} \quad \text{and } \overrightarrow{b} = \begin{pmatrix} b_1 \\ b_2 \end{pmatrix} = \begin{pmatrix} 7 \\ 1 \end{pmatrix}.$$

Then by (4.21),

$$A_1 = \begin{pmatrix} b_1 & a_{12} \\ b_2 & a_{22} \end{pmatrix} = \begin{pmatrix} 7 & -3 \\ 1 & 5 \end{pmatrix} \quad \text{and } A_2 = \begin{pmatrix} a_{11} & b_1 \\ a_{21} & b_2 \end{pmatrix} = \begin{pmatrix} 2 & 7 \\ 1 & 1 \end{pmatrix}.$$

Hence, we have

$$\begin{aligned} |A| &= \begin{vmatrix} 2 & -3 \\ 1 & 5 \end{vmatrix} = 10 + 3 = 13, \\ |A_1| &= \begin{vmatrix} 7 & -3 \\ 1 & 5 \end{vmatrix} = 35 + 3 = 38, \\ |A_2| &= \begin{vmatrix} 2 & 7 \\ 1 & 1 \end{vmatrix} = 2 - 7 = -5. \end{aligned}$$

By (4.22),

$$x_1 = \frac{|A_1|}{|A|} = \frac{38}{13} \quad \text{and } x_2 = \frac{|A_2|}{|A|} = -\frac{5}{13}.$$

Hence, $(\frac{38}{13}, -\frac{5}{13})$ is a solution of the system. □

Example 4.10.3. Use Cramer's rule to solve

$$\begin{cases} x_1 + 2x_3 = 6 \\ -3x_1 + 4x_2 + 6x_3 = 30 \\ -x_1 - 2x_2 + 3x_3 = 8. \end{cases} \tag{4.23}$$

Solution. Let

$$A = \begin{pmatrix} a_{11} & a_{12} & a_{13} \\ a_{21} & a_{22} & a_{23} \\ a_{31} & a_{32} & a_{33} \end{pmatrix} = \begin{pmatrix} 1 & 0 & 2 \\ -3 & 4 & 6 \\ -1 & -2 & 3 \end{pmatrix}$$

and

$$\overrightarrow{b} = \begin{pmatrix} b_1 \\ b_2 \\ b_3 \end{pmatrix} = \begin{pmatrix} 6 \\ 30 \\ 8 \end{pmatrix}.$$

Then by (4.21),

$$A_1 = \begin{pmatrix} b_1 & a_{12} & a_{13} \\ b_2 & a_{22} & a_{23} \\ b_3 & a_{32} & a_{33} \end{pmatrix} = \begin{pmatrix} 6 & 0 & 2 \\ 30 & 4 & 6 \\ 8 & -2 & 3 \end{pmatrix}$$

$$A_2 = \begin{pmatrix} a_{11} & b_1 & a_{13} \\ a_{21} & b_2 & a_{23} \\ a_{31} & b_3 & a_{33} \end{pmatrix} = \begin{pmatrix} 1 & 6 & 2 \\ -3 & 30 & 6 \\ -1 & 8 & 3 \end{pmatrix}$$

$$A_3 = \begin{pmatrix} a_{11} & a_{12} & b_1 \\ a_{21} & a_{22} & b_2 \\ a_{31} & a_{32} & b_3 \end{pmatrix} = \begin{pmatrix} 1 & 0 & 6 \\ -3 & 4 & 30 \\ -1 & -2 & 8 \end{pmatrix}.$$

By computation, we have

$$|A| = 44, \quad |A_1| = -40, \quad |A_2| = 72 \quad \text{and } |A_3| = 152.$$

By (4.22),

$$\begin{aligned} x_1 &= \frac{|A_1|}{|A|} = \frac{-40}{44} = -\frac{10}{11}, \\ x_2 &= \frac{|A_2|}{|A|} = \frac{72}{44} = \frac{18}{11}, \\ x_3 &= \frac{|A_3|}{|A|} = \frac{152}{44} = \frac{38}{11}. \end{aligned}$$

Hence, $(-\frac{10}{11}, \frac{18}{11}, \frac{38}{11})$ is a solution of the system. □

Example 4.10.4.

$$\begin{cases} x_1 + x_2 + x_3 = 0 \\ 2x_1 - x_2 - x_3 = 1 \\ x_2 + 2x_3 = 2 \end{cases}$$

Solution. Let

$$A = \begin{pmatrix} 1 & 1 & 1 \\ 2 & -1 & -1 \\ 1 & 0 & 2 \end{pmatrix}$$

Then

$$A_1 = \begin{pmatrix} 0 & 1 & 1 \\ 1 & -1 & -1 \\ 2 & 0 & 2 \end{pmatrix}, A_2 = \begin{pmatrix} 1 & 0 & 1 \\ 2 & 1 & -1 \\ 1 & 2 & 2 \end{pmatrix} \text{ and } A_3 = \begin{pmatrix} 1 & 1 & 0 \\ 2 & -1 & 1 \\ 1 & 0 & 2 \end{pmatrix}.$$

By computation, we obtain

$$|A| = -3, |A_1| = -1, |A_2| = 8 \quad \text{and } |A_3| = -7.$$

By (4.22),

$$x_1 = \frac{1}{3}, \quad x_2 = -\frac{8}{3} \quad \text{and } x_3 = \frac{7}{3}.$$

Hence, $(\frac{1}{3}, -\frac{8}{3}, \frac{7}{3})$ is a solution of the system. □

Exercises

1. Solve the following system by using Cramer's rule.

   a) $\begin{cases} x - 3y = 6 \\ 2x + 5y = -1. \end{cases}$

   b) $\begin{cases} x_1 + 2x_3 = 2 \\ -2x_1 + 3x_2 + x_3 = 16 \\ -x_1 - 2x_2 + 4x_3 = 5. \end{cases}$

   c) $\begin{cases} x_1 + x_2 - x_3 = 1 \\ 2x_1 + x_2 - x_3 = 0 \\ x_2 - 2x_3 = -2. \end{cases}$

Chapter 5

Vectors in $\mathbb{R}^3$

5.1 Norm, angle, orthogonal and projection

5.1.1 Norm of a vector

Let $\overrightarrow{u} = (u_1, u_2, u_3) \in \mathbb{R}^3$. Then we define the norm of the vector $\overrightarrow{u}$ as follows:

$$\|\overrightarrow{u}\| = \sqrt{u_1^2 + u_2^2 + u_3^2}. \tag{5.1}$$

Geometrically, the norm $\|\overrightarrow{u}\|$ represents the length of the vector $\overrightarrow{u}$.

Example 5.1.1. Let $\overrightarrow{u} = (1, 3, -2)$. Find $\|\overrightarrow{u}\|$.

Solution. $\|\overrightarrow{u}\| = \sqrt{1^2 + 3^2 + (-2)^2} = \sqrt{14}$. □

Given two points $P_1(x_1, x_2, x_3)$ and $P_2(y_1, y_2, y_3)$ in $\mathbb{R}^3$. Then

$$\overrightarrow{P_1P_2} = (y_1 - x_1, y_2 - x_2, y_3 - x_3). \tag{5.2}$$

We denote by $d(P_1P_2)$ the distance of the two points P_1 and P_2. Then $d(P_1P_2)$ equals the norm of the vector $\overline{P_1P_2}$, that is,

$$d(P_1P_2) = \|\overrightarrow{P_1P_2}\| = \sqrt{(y_1 - x_1)^2 + (y_2 - x_2)^2 + (y_3 - x_3)^2}. \tag{5.3}$$

Example 5.1.2. Find the distance between $P_1(2, -1, 5)$ and $P_2(4, -3, 1)$.

Solution. $d(P_1P_2) = \|\overrightarrow{P_1P_2}\| = \sqrt{(4-2)^2 + (-3+1)^2 + (1-5)^2} = 2\sqrt{6}$. □

5.1.2 Angle between two vectors

Let $\overrightarrow{u}, \overrightarrow{v} \in \mathbb{R}^3$. We denote by θ the angle between the two vectors $\overrightarrow{u}$ and $\overrightarrow{v}$. Then $0 \leq \theta \leq \pi$.

The following result provides a formula to compute the angle θ.

Theorem 5.1.3. $\cos\theta = \dfrac{\overrightarrow{u} \cdot \overrightarrow{v}}{\|\overrightarrow{u}\|\|\overrightarrow{v}\|}$.

Example 5.1.4. Let $\overrightarrow{u} = (2, -1, 1)$ and $\overrightarrow{v} = (1, 1, 2)$. Find the angle θ between $\overrightarrow{u}$ and $\overrightarrow{v}$

Solution. Since

$$\begin{aligned} \overrightarrow{u} \cdot \overrightarrow{v} &= 2 \times 1 + (-1) \times 1 + 1 \times 2 = 3, \\ \|\overrightarrow{u}\| &= \sqrt{2^2 + (-1)^2 + 1^2} = \sqrt{6}, \\ \|\overrightarrow{v}\| &= \sqrt{1^2 + 1^2 + 4} = \sqrt{6}, \end{aligned}$$

we have by Theorem 5.1.3,

$$\cos\theta = \frac{\overrightarrow{u} \cdot \overrightarrow{v}}{\|\overrightarrow{u}\|\|\overrightarrow{v}\|} = \frac{3}{\sqrt{6}\sqrt{6}} = \frac{1}{2}.$$

Since $0 \leq \theta \leq \pi$, we have $\theta = \frac{\pi}{3}$. □

Example 5.1.5. Let $\overrightarrow{u} = (a, a, a)$, where $a > 0$ is given and $\overrightarrow{v} = (a, 0, 0)$. Find the angle θ between $\overrightarrow{u}$ and $\overrightarrow{v}$.

Solution. By Theorem 5.1.3, we have

$$\cos\theta = \frac{\overrightarrow{u} \cdot \overrightarrow{v}}{\|\overrightarrow{u}\|\|\overrightarrow{v}\|} = \frac{a^2}{\sqrt{3a^2} \cdot \sqrt{a^2}} = \frac{1}{\sqrt{3}} = \frac{\sqrt{3}}{3}.$$

Hence, $\theta = \arccos\frac{\sqrt{3}}{3} \approx 54.74^o$. □

5.1.3 Orthogonal vectors

Let $\overrightarrow{u}, \overrightarrow{v} \in \mathbb{R}^3$ and let θ be the angle between $\overrightarrow{u}$ and $\overrightarrow{v}$. Then $\overrightarrow{u} \perp \overrightarrow{v}$ if and only if $\theta = \pi/2$. By Theorem 5.1.3,

$$\overrightarrow{u} \perp \overrightarrow{v} \quad \text{if and only if } \overrightarrow{u} \cdot \overrightarrow{v} = 0. \tag{5.4}$$

Example 5.1.6. Let $\overrightarrow{u} = (-2, 3, 1)$ and $\overrightarrow{v} = (1, 2, -4)$. Show $\overrightarrow{u} \perp \overrightarrow{v}$.

Solution. Since $\overrightarrow{u} \cdot \overrightarrow{v} = -2 + 6 - 4 = 0$, by (5.4), we have $\overrightarrow{u} \perp \overrightarrow{v}$. □

The following result gives a property of two orthogonal vectors.

Theorem 5.1.7. *Let $\overrightarrow{u}, \overrightarrow{v} \in \mathbb{R}^3$. Assume that $\overrightarrow{u} \perp \overrightarrow{v}$. Then*

$$\|\overrightarrow{u} + \overrightarrow{v}\|^2 = \|\overrightarrow{u}\|^2 + \|\overrightarrow{v}\|^2.$$

5.1.4 Projection in $\mathbb{R}^3$

Let $\overrightarrow{u}$, $\overrightarrow{a} \in \mathbb{R}^3$. We define the projection of $\overrightarrow{u}$ to $\overrightarrow{a}$ by

$$\text{proj}_{\overrightarrow{a}}\, \overrightarrow{u} = \left(\frac{\overrightarrow{u} \cdot \overrightarrow{a}}{\|\overrightarrow{a}\|^2}\right)\overrightarrow{a}. \tag{5.5}$$

Example 5.1.8. Let $\overrightarrow{u} = (2, -1, 3)$ and $\overrightarrow{a} = (4, -1, 2)$. Find $\text{proj}_{\overrightarrow{a}}\, \overrightarrow{u}$.

Solution. Since $\overrightarrow{u} \cdot \overrightarrow{a} = 2\times 4 + (-1)(-1) + 3\times 2 = 15$ and $\|\overrightarrow{a}\|^2 = 16+1+4 = 21$, by (5.5) we obtain

$$\text{proj}_{\overrightarrow{a}}\, \overrightarrow{u} = \frac{15}{21}(4, -1, 2) = \frac{5}{7}(4, -1, 2) = \left(\frac{20}{7}, -\frac{5}{7}, \frac{10}{7}\right).$$

□

Exercises

1. Find $\|\overrightarrow{u}\|$.

   (1) $\overrightarrow{u} = (1, 0, -2)$; (2) $\overrightarrow{u} = (1, 1, -2)$; (3) $\overrightarrow{u} = (2, 2, -2)$.

2. Find the distance between P_1 and P_2.

   (1) $P_1(1, -1, 5)$ and $P_2(1, -3, 1)$; (2) $P_1(0, -1, 2)$ and $P_2(1, -3, 1)$

3. Find the cosine of the angle θ between $\overrightarrow{u}$ and $\overrightarrow{v}$.

   (1) $\overrightarrow{u} = (1, -1, 1)$ and $\overrightarrow{v} = (1, 1, -2)$;

   (2) $\overrightarrow{u} = (1, 0, 1)$ and $\overrightarrow{v} = (-1, -1, -1)$.

4. Find $\text{proj}_{\overrightarrow{a}}\, \overrightarrow{u}$.

   (1) $\overrightarrow{u} = (1, -1, 2)$ and $\overrightarrow{a} = (2, -1, 1)$;

   (2) $\overrightarrow{u} = (1, 0, -1)$ and $\overrightarrow{a} = (-1, -2, 2)$.

5.2 The cross product of two vectors in $\mathbb{R}^3$

Let $\overrightarrow{u} = (u_1, u_2, u_3)$ and $\overrightarrow{v} = (v_1, v_2, v_3)$. Then we define the cross product of $\overrightarrow{u}$ and $\overrightarrow{v}$ as a vector in $\mathbb{R}^3$:

$$\overrightarrow{u} \times \overrightarrow{v} = (u_2v_3 - u_3v_2, -(u_1v_3 - u_3v_1), u_1v_2 - u_2v_1). \tag{5.6}$$

We can rewrite $\overrightarrow{u} \times \overrightarrow{v}$ as follows:

$$\begin{aligned}
\overrightarrow{u} \times \overrightarrow{u} &= \left(\begin{vmatrix} u_2 & u_3 \\ v_2 & v_3 \end{vmatrix}, - \begin{vmatrix} u_1 & u_3 \\ v_1 & v_3 \end{vmatrix}, \begin{vmatrix} u_1 & u_2 \\ v_1 & v_2 \end{vmatrix} \right) & (5.7) \\
&= \begin{vmatrix} u_2 & u_3 \\ v_2 & v_3 \end{vmatrix} \overrightarrow{i} - \begin{vmatrix} u_1 & u_3 \\ v_1 & v_3 \end{vmatrix} \overrightarrow{j} + \begin{vmatrix} u_1 & u_2 \\ v_1 & v_2 \end{vmatrix} \overrightarrow{k} & (5.8) \\
&= \begin{vmatrix} \overrightarrow{i} & \overrightarrow{j} & \overrightarrow{k} \\ u_1 & u_2 & u_3 \\ v_1 & v_2 & v_3 \end{vmatrix}. & (5.9)
\end{aligned}$$

where $\overrightarrow{i} = (1, 0, 0)$, $\overrightarrow{j} = (0, 1, 0)$ and $\overrightarrow{k} = (0, 0, 1)$.

Theorem 5.2.1. *The cross product of $\overrightarrow{u}$ and $\overrightarrow{v}$ has the following properties:*

(1) $\overrightarrow{u} \times \overrightarrow{v} \perp \overrightarrow{u}$.

(2) $\overrightarrow{u} \times \overrightarrow{v} \perp \overrightarrow{v}$.

Solution. We only show (1). By (5.7), we have

$$\begin{aligned}
(\overrightarrow{u} \times \overrightarrow{v}) \cdot \overrightarrow{u} &= \begin{vmatrix} u_2 & u_3 \\ v_2 & v_3 \end{vmatrix} u_1 - \begin{vmatrix} u_1 & u_3 \\ v_1 & v_3 \end{vmatrix} u_2 + \begin{vmatrix} u_1 & u_2 \\ v_1 & v_2 \end{vmatrix} u_3 \\
&= \begin{vmatrix} u_1 & u_2 & u_3 \\ u_1 & u_2 & u_3 \\ v_1 & v_2 & v_3 \end{vmatrix} = 0.
\end{aligned}$$

Hence, $\overrightarrow{u} \times \overrightarrow{v} \perp \overrightarrow{u}$. □

Example 5.2.2. Let $\overrightarrow{u} = (1, 2, -2)$ and $\overrightarrow{v} = (3, 0, 1)$. Find $\overrightarrow{u} \times \overrightarrow{v}$ and verify that $\overrightarrow{u} \times \overrightarrow{v} \perp \overrightarrow{u}$ and $\overrightarrow{u} \times \overrightarrow{v} \perp \overrightarrow{v}$.

Solution.

$$\begin{aligned}
\overrightarrow{u} \times \overrightarrow{v} &= \begin{vmatrix} \overrightarrow{i} & \overrightarrow{j} & \overrightarrow{k} \\ 1 & 2 & -2 \\ 3 & 0 & 1 \end{vmatrix} = \begin{vmatrix} 2 & -2 \\ 0 & 1 \end{vmatrix} \overrightarrow{i} - \begin{vmatrix} 1 & -2 \\ 3 & 1 \end{vmatrix} \overrightarrow{j} + \begin{vmatrix} 1 & 2 \\ 3 & 0 \end{vmatrix} \overrightarrow{k} \\
&= 2\overrightarrow{i} - 7\overrightarrow{j} - 6\overrightarrow{k} = (2, -7, 6).
\end{aligned}$$

$$(\overrightarrow{u} \times \overrightarrow{v}) \cdot \overrightarrow{u} = (2, -7, -6) \begin{pmatrix} 1 \\ 2 \\ -2 \end{pmatrix} = 0.$$

and

$$(\overrightarrow{u} \times \overrightarrow{v}) \cdot \overrightarrow{v} = (2, -7, -6) \begin{pmatrix} 3 \\ 0 \\ 1 \end{pmatrix} = 0.$$

□

Theorem 5.2.3. (1) *Let $\overrightarrow{u} = (u_1, u_2, u_3)$ and $\overrightarrow{v} = (v_1, v_2, v_3)$. Then the area of the parallelogram determined by $\overrightarrow{u}$ and $\overrightarrow{v}$ is*

$$\|\overrightarrow{u} \times \overrightarrow{v}\|.$$

(2) *Let $\overrightarrow{u} = (u_1, u_2)$ and $\overrightarrow{v} = (v_1, v_2)$. Then the area of the parallelogram determined by $\overrightarrow{u}$ and $\overrightarrow{v}$ is*

$$\left|\det \begin{vmatrix} u_1 & u_2 \\ v_1 & v_2 \end{vmatrix}\right| = |u_1 v_2 - u_2 v_1|.$$

Example 5.2.4. Find the area of the parallelogram determined by $\overrightarrow{u} = (1, -2, 0)$ and $\overrightarrow{v} = (-5, 0, 2)$.

Solution. By Theorem 5.2.3 (1), the area of the parallelogram is

$$\|\overrightarrow{u} \times \overrightarrow{v}\| = \|(-4, -2, -10)\| = \sqrt{(-4)^2 + (-2)^2 + (-10)^2} = \sqrt{120} = 2\sqrt{30}.$$

□

Example 5.2.5. Find the area of the parallelogram determined by $\overrightarrow{u} = (1, 2)$ and $\overrightarrow{v} = (0, -1)$

Solution. By Theorem 5.2.3 (2), the area of the parallelogram is

$$A = \left|\det \begin{vmatrix} 1 & 2 \\ 0 & -1 \end{vmatrix}\right| = |1 \times (-1) - 2 \times 0| = 1.$$

□

2. Scalar triple product

Let $\overrightarrow{u} = (u_1, u_2, u_3)$, $\overrightarrow{v} = (v_1, v_2, v_3)$ and $\overrightarrow{w} = (w_1, w_2, w_3)$. Then

$$\overrightarrow{u} \cdot (\overrightarrow{v} \times \overrightarrow{w}) \tag{5.10}$$

is a scalar triple product.

Example 5.2.6. Let $\overrightarrow{u} = (3, -2, -5)$, $\overrightarrow{v} = (1, 4, -4)$ and $\overrightarrow{w} = (0, 3, 2)$. Find $\overrightarrow{u} \cdot (\overrightarrow{v} \times \overrightarrow{w})$.

Solution. By (5.6), we have

$$\begin{aligned} \overrightarrow{v} \times \overrightarrow{w} &= \begin{vmatrix} \overrightarrow{i} & \overrightarrow{j} & \overrightarrow{k} \\ 1 & 4 & -4 \\ 0 & 3 & 2 \end{vmatrix} = \begin{vmatrix} 4 & -4 \\ 3 & 2 \end{vmatrix} \overrightarrow{i} - \begin{vmatrix} 1 & -4 \\ 0 & 2 \end{vmatrix} \overrightarrow{j} + \begin{vmatrix} 1 & 4 \\ 0 & 3 \end{vmatrix} \overrightarrow{k} \\ &= (8 + 12)\overrightarrow{i} - (2 - 0)\overrightarrow{j} + (3 - 0)\overrightarrow{k} = 20\overrightarrow{i} - 2\overrightarrow{j} + 3\overrightarrow{k} \\ &= (20, -2, 3). \end{aligned}$$

Hence,

$$\overrightarrow{u} \cdot (\overrightarrow{v} \times \overrightarrow{w}) = (3, -2, -5) \begin{pmatrix} 20 \\ -2 \\ 3 \end{pmatrix} = 3 \times 20 + 4 - 15 = 64 - 15 = 49.$$

□

The scalar triple product can be computed by a 3×3 determinant.

Theorem 5.2.7. $\overrightarrow{u} \cdot (\overrightarrow{v} \times \overrightarrow{w}) = \begin{vmatrix} u_1 & u_2 & u_3 \\ v_1 & v_2 & v_3 \\ w_1 & w_2 & w_3 \end{vmatrix}.$

Example 5.2.8. Use Theorem 5.2.7 to find $\overrightarrow{u} \cdot (\overrightarrow{v} \times \overrightarrow{w})$ in Example 5.2.6.

Solution. $\overrightarrow{u} \cdot (\overrightarrow{v} \times \overrightarrow{w}) = \begin{vmatrix} 3 & -2 & -5 \\ 1 & 4 & -4 \\ 0 & 3 & 2 \end{vmatrix} = 49.$ □

Theorem 5.2.9. *Let $\overrightarrow{u} = (u_1, u_2, u_3)$, $\overrightarrow{v} = (v_1, v_2, v_3)$, $\overrightarrow{w} = (w_1, w_2, w_3)$ be three vectors which are not in the same plane. Then the volume of the parallelpiped determined by the three vectors is given by*

$$V = |\overrightarrow{u} \cdot (\overrightarrow{v} \times \overrightarrow{w})|$$

Example 5.2.10. Let $\overrightarrow{u} = (3, -2, -5)$, $\overrightarrow{v} = (1, 4, -4)$ and $\overrightarrow{w} = (0, -3, -2)$. Find the volume of the parallelpiped determined by $\overrightarrow{u}$, $\overrightarrow{v}$ and $\overrightarrow{w}$.

Solution. By computation, we obtain

$$\overrightarrow{v} \times \overrightarrow{w} = (-20, 2, -3).$$

Hence,

$$\begin{aligned} \overrightarrow{u} \cdot (\overrightarrow{v} \times \overrightarrow{w}) &= (3, -2, -5) \begin{pmatrix} -20 \\ 2 \\ -3 \end{pmatrix} = (3)(-20) + (-2)(2) + (-5)(-3) \\ &= -60 - 4 + 15 = -49. \end{aligned}$$

By Theorem 5.2.9, we obtain $V = |\overrightarrow{u} \cdot (\overrightarrow{v} \times \overrightarrow{w})| = |-49| = 49.$ □

Exercises

1. Find the cross product $\overrightarrow{u} \times \overrightarrow{v}$.

   (1) $\overrightarrow{u} = (1, 2, 3)$; $\overrightarrow{v} = (1, 0, 1)$, (2) $\overrightarrow{u} = (1, 0, -3)$; $\overrightarrow{v} = (-1, 0, -2)$
   (3) $\overrightarrow{u} = (0, 2, 1)$; $\overrightarrow{v} = (-5, 0, 1)$, (4) $\overrightarrow{u} = (1, 1, -1)$; $\overrightarrow{v} = (-1, 1, 0)$.

2. Find the area of the parallelogram determined by $\overrightarrow{u}$ and $\overrightarrow{v}$.

   (1) $\overrightarrow{u} = (1,0,1); \overrightarrow{v} = (-1,0,1)$, (2) $\overrightarrow{u} = (1,0,0); \overrightarrow{v} = (-1,1,-2)$
   (3) $\overrightarrow{u} = (0,2,-1); \overrightarrow{v} = (-5,1,1)$, (4) $\overrightarrow{u} = (1,-1,-1); \overrightarrow{v} = (-1,1,2)$.

3. Find the area of the parallelogram determined by $\overrightarrow{u}$ and $\overrightarrow{v}$.

   (1) $\overrightarrow{u} = (1,0); \overrightarrow{v} = (0,-1)$, (2) $\overrightarrow{u} = (1,0); \overrightarrow{v} = (-1,1)$
   (3) $\overrightarrow{u} = (0,2); \overrightarrow{v} = (1,1)$, (4) $\overrightarrow{u} = (1,-1); \overrightarrow{v} = (-1,2)$.

4. Find the volume of the parallelpiped determined by $\overrightarrow{u}$, $\overrightarrow{u}$ and $\overrightarrow{u}$.

   (1) $\overrightarrow{u} = (1,-1,0)$, $\overrightarrow{v} = (-1,0,2)$ and $\overrightarrow{w} = (0,-1,-1)$.

   (2) $\overrightarrow{u} = (-1,-1,0)$, $\overrightarrow{v} = (-1,1,-2)$ and $\overrightarrow{w} = (-1,1,1)$.

5.3 Planes and lines in $\mathbb{R}^3$

5.3.1 Planes in $\mathbb{R}^3$

1. Point-normal form

Given a vector $\overrightarrow{n} = (a,b,c)$ and a point $P_0(x_0,y_0,z_0)$. We need to find an equation of the plane that is orthogonal to $\overrightarrow{n}$ and passes through P_0.

The vector $\overrightarrow{n}$ is called the normal vector of the plane and the equation of the plane is called the point-normal form.

Before we establish the equation of the plane, we clarify some facts:

(P_1) A vector $\overrightarrow{n}$ is orthogonal to a plane if and only if it is orthogonal to any vector that is parallel to the plane.

(P_2) If both $\overrightarrow{u}$ and $\overrightarrow{v}$ are parallel to the plane and $\overrightarrow{u}$ is not parallel to $\overrightarrow{v}$, then the cross product $\overrightarrow{u} \times \overrightarrow{v}$ is orthogonal to a plane.

(P_3) If $P_1(x_1,y_1,z_1)$ and $P_2(x_2,y_2,z_2)$ are in the plane, then the vector $\overrightarrow{P_1P_2}$ is parallel to the plane.

Now, we take a point $P(x,y,z)$ in the plane. Then P_0 and P are in the plane. By property (P_3), the vector $\overrightarrow{P_0P}$ is parallel to the plane. By the assumption, the plane is orthogonal to $\overrightarrow{n}$. Hence, $\overrightarrow{n}$ is orthogonal to the plane. By property (P_1), $\overrightarrow{n}$ is orthogonal to $\overrightarrow{P_0P}$. This implies that

$$\overrightarrow{n} \cdot \overrightarrow{P_0P} = 0.$$

Since $\overrightarrow{n} = (a,b,c)$ and $\overrightarrow{P_0P} = (x-x_0, y-y_0, z-z_0)$, we obtain

$$\overrightarrow{n} \cdot \overrightarrow{P_0P} = a(x-x_0) + b(y-y_0) + c(z-z_0) = 0.$$

This implies

$$ax + by + cz = ax_0 + by_0 + cz_0.$$

Hence, the equation of the plane is

$$ax + by + cz = d, \tag{5.11}$$

where $d = ax_0 + by_0 + cz_0$.

Example 5.3.1. Find the equation of the plane passing through $(3, -1, 7)$ and perpendicular to $\overrightarrow{n} = (4, 2, -5)$.

Solution. By (5.11), the equation is

$$4x + 2y - 5z = 4(3) + 2(-1) + (-5)(7) = -25.$$

□

Example 5.3.2. Find the normal vector of the plane $4x + 2y - 5z = 3$.

Solution. By (5.11), $\overrightarrow{n} = (4, 2, -5)$. □

Example 5.3.3. Find the equation of the plane passing through the following three points:

$$P_1(1, 2, -1), \quad P_2(2, 3, 1), \quad P_3(3, -1, 2).$$

Solution. **Method 1.** Let $\overrightarrow{u} = \overrightarrow{P_1P_2}$ and $\overrightarrow{v} = \overrightarrow{P_1P_3}$. Then

$$\overrightarrow{u} = \overrightarrow{P_1P_2} = (x_2 - x_1, y_2 - y_1, z_2 - z_1) = (1, 1, 2)$$

and

$$\overrightarrow{v} = \overrightarrow{P_1P_3} = (x_3 - x_1, y_3 - y_1, z_3 - z_1) = (2, -3, 3).$$

Moreover, $\overrightarrow{u}$ is not parallel to $\overrightarrow{v}$ since the corresponding coordinates are not proportional. By (P_3), each of $\overrightarrow{u}$ and $\overrightarrow{v}$ is parallel to the plane. Let $\overrightarrow{n} = \overrightarrow{u} \times \overrightarrow{v}$. By computation, we obtain

$$\overrightarrow{n} = \overrightarrow{u} \times \overrightarrow{v} = (-9, -1, 5).$$

Let $P_0(x_0, y_0, z_0) = P_1(1, 2, -1)$. By (5.11), we have

$$9z + y - 5z = 16.$$

Method 2. Assume that the equation of the required plane is

$$ax + by + cz = d. \tag{5.12}$$

. We need to determine a, b, c, d in order to get the exact equation. Since P_1, P_2, P_3 are in the plane, so their coordinates satisfy (5.12), that is,

$$\begin{cases} a + 2b - c = d \\ 2a + 3b + c = d \\ 3a - b + 2c = d. \end{cases}$$

Now we solve the above system for a, b, c and treat d as a constant.

$$(A|\overrightarrow{b}) = \left(\begin{array}{ccc|c} 1 & 2 & -1 & d \\ 2 & 3 & 1 & d \\ 3 & -1 & 2 & d \end{array}\right) \xrightarrow[R_1(-3)+R_3]{R_1(-2)+R_2} \left(\begin{array}{ccc|c} 1 & 2 & -1 & d \\ 0 & -1 & 3 & -d \\ 0 & -7 & 5 & -2d \end{array}\right)$$

$$\xrightarrow{R_2(-1)} \left(\begin{array}{ccc|c} 1 & 2 & -1 & d \\ 0 & 1 & -3 & d \\ 0 & -7 & 5 & -2d \end{array}\right) \xrightarrow[R_2(7)+R_3]{R_2(-2)+R_1}$$

$$\left(\begin{array}{ccc|c} 1 & 0 & 5 & -d \\ 0 & 1 & -3 & d \\ 0 & 0 & -16 & 5d \end{array}\right) \xrightarrow{R_3(-\frac{1}{16})} \left(\begin{array}{ccc|c} 1 & 0 & 5 & -d \\ 0 & 1 & -3 & d \\ 0 & 0 & 1 & -\frac{5}{16}d \end{array}\right)$$

$$\xrightarrow[R_3(3)+R_2]{R_3(-5)+R_1} \left(\begin{array}{ccc|c} 1 & 0 & 0 & \frac{9}{16}d \\ 0 & 1 & 0 & \frac{1}{16}d \\ 0 & 0 & 1 & -\frac{5}{16}d \end{array}\right).$$

Hence, $a = \frac{9}{16}d$, $b = \frac{1}{16}d$ and $c = -\frac{5}{16}d$. By (5.12), we obtain

$$(\frac{9}{16}d)x + (\frac{d}{16})y - (\frac{5}{16}d)z = d.$$

Taking $d = 1$ and simplifying implies

$$9x + y - 5z = 16.$$

□

2. Distance between a point and a plane

The distance D between a point $P_0(x_0, y_0, z_0)$ and the plane $ax + by + cz + d = 0$ is

$$D = \frac{|ax_0 + by_0 + cz_0 + d|}{\sqrt{a^2 + b^2 + c^2}}. \tag{5.13}$$

Example 5.3.4. Find the distance between the point $P(1, -4, -3)$ and the plane $2x - 3y + 6z = -1$.

Solution. By (5.13), we have

$$D = \frac{|ax_0 + by_0 + cz_0 + d|}{\sqrt{a^2 + b^2 + c^2}} = \frac{|2(1) - 3(-4) + 6(-3) + 1|}{\sqrt{2^2 + (-3)^2 + 6^2}} = \frac{|-3|}{7} = \frac{3}{7}.$$

□

Example 5.3.5. Find the distance between the following two parallel planes

$$\begin{cases} x + 2y = 2z = 3 - - - - - - - - - - - -(1) \\ 2x + 4y - 4z = 7 - - - - - - - - - - - -(2) \end{cases}$$

Solution. In (1), let $y = z = 0$. Then we have $x = 3$. So $P_0(3, 0, 0)$ is a point in the plane (1). The distance between the point $P_0(3, 0, 0)$ and the plane (2) equals the distance between the two parallel planes (1) and (2). By (5.13), we have

$$D = \frac{|2 \times 3 + 0 + 0 - 7|}{\sqrt{2^2 + 4^2 + (-4)^2}} = \frac{1}{6}.$$

□

5.3.2 Equations of lines in $\mathbb{R}^3$

Let $\overrightarrow{u}$, $\overrightarrow{v}$ be vectors in $\mathbb{R}^3$. Recall that $\overrightarrow{u}$ is parallel to $\overrightarrow{v}$ if and only if there exists a constant $t \in \mathbb{R}$ such that

$$\overrightarrow{u} = t\overrightarrow{v}.$$

This shows that the corresponding coordinates of $\overrightarrow{u}$ and $t\overrightarrow{v}$ are proportional.

Given a point $P_0(x_0, y_0, z_0)$ and a vector $\overrightarrow{v} = (a, b, c)$. We find the equation of the line that passes through $P_0(x_0, y_0, z_0)$ and is parallel to the vector $\overrightarrow{v}$.

Let $P(x, y, z)$ be any point on the line. Then $\overrightarrow{P_0P} = (x - x_0, y - y_0, z - z_0)$ and $\overrightarrow{P_0P}$ is parallel to $\overrightarrow{v}$. Hence, there exists a constant $t \in \mathbb{R}$ such that

$$\overrightarrow{P_0P} = t\overrightarrow{v}.$$

This implies

$$(x - x_0, y - y_0, z - z_0) = t(a, b, c)$$

This implies

$$\begin{cases} x - x_0 = ta, \\ y - y_0 = tb, \qquad -\infty < t < \infty \\ z - z_0 = tc \end{cases}$$

and

$$\begin{cases} x = x_0 + ta, \\ y = y_0 + tb, \qquad -\infty < t < \infty \\ z = z_0 + tc. \end{cases} \tag{5.14}$$

(5.14) is said to be a parametric equation, where t is a parameter.

Example 5.3.6. Find the parametric equation of the line passing through the point $P_0(1,2,-3)$ that is parallel to the vector $\overrightarrow{v}=(4,5,-7)$.

Solution. By (5.14), we obtain

$$\begin{cases} x = 1+4t, \\ y = 2+5t, \\ z = -3-7t. \end{cases} \qquad -\infty < t < \infty,$$

□

Example 5.3.7. (a) Find the equation of the line that passes through $P_1(2,4,-1)$ and $P_2(5,0,7)$.

(b) Where does the line intersect the xy-plane.

Solution. (*a*) Let $\overrightarrow{v}=\overrightarrow{P_1P_2}=(3,-4,8)$ and $P_0=P_1(2,4,-1)$. By (5.14), we have

$$\begin{cases} x = 2+3t, \\ y = 4-4t, \\ z = -1+8t. \end{cases} \qquad -\infty < t < \infty,$$

(b) Since the line intersects the xy-plane, $z=0$. Let $z=-1+8t=0$. Then $t=\frac{1}{8}$. Hence,

$$x = 2+3(\frac{1}{8}) = \frac{19}{8} \quad \text{and } y = 4-4(\tfrac{1}{8}) = \tfrac{7}{2}.$$

Hence, the intersection point is $(\frac{10}{8},\frac{7}{2},0)$ □

Example 5.3.8. Find the parametric equation of the line of intersection of the following two planes

$$\begin{cases} 3x+2y-4z-6=0 \\ x-3y-2z-4=0. \end{cases}$$

Solution. Solving (1) and (2), we obtain the parametric equation

$$\begin{cases} x = \frac{26}{11}+\frac{16}{11}t, \\ y = -\frac{6}{11}-\frac{2}{11}t, \\ z = t. \end{cases} \qquad -\infty < z < \infty,$$

□

Example 5.3.9. Find an equation for the line through $P_0(2, -3, 0)$ that is parallel to the planes $2x + 2y + z = 2$ and $x - 3y = 5$.

Solution. We need to find the vector which is parallel to the line. Since the line is parallel to the two planes, the vector we are looking for is parallel to the two planes. By property (P_1), the vector is orthogonal to the normal vector of each of the two planes. Hence, we can choose the vector as the cross product of the two normal vectors. Hence, assume that $\overrightarrow{n_1}$ and $\overrightarrow{n_2}$ are the normal vectors of the planes, respectively. Then $\overrightarrow{n_1} = (2, 2, 1)$ and $\overrightarrow{n_2} = (1, -3, 0)$. Let $\overrightarrow{v} = \overrightarrow{n_1} \times \overrightarrow{n_2}$. Then by computation, we have

$$\overrightarrow{v} = (2, 2, 1) \times (1, -3, 0) = (3, 1, -8).$$

Since the line passes through $P_0 = (2, -3, 0)$, it follows from (5.14) that

$$\begin{cases} x = 2 + 3t, \\ y = -3 + t, \\ z = -8t, \end{cases} \qquad -\infty < t < \infty.$$

□

Exercises

1. Find an equation for the plane through $P_0(1, 1, 3)$ that is perpendicular to the line
$$\begin{cases} x = 2 - 3t \\ y = 1 + t \\ z = 2t \end{cases}$$

2. Find and equation for the plane through $P_0(2, 7, -1)$ that is parallel to the plane $4x - y + 3z = 3$.

3. Find and equation for the plane that contains the line $x = 3 + t$, $y = 5$, $z = t + 2t$, and is perpendicular to the plane $x + y + z = 4$.

4. Find an equation for the plane through $P_0(1, 4, 4)$ that contains the line of intersection of the planes
$$\begin{cases} x - y + 3z = 5 \\ 2x + 2y + 7z = 0 \end{cases}$$

5. Find an equation for the plane containing the line $x = 3 + 6t$, $y = 4$, $z = t$ and that is parallel to the line of intersection of the plane $2x + y + z = 1$ and $x - 2y + 3z = 2$.

Appendix A

Solutions to the Problems

A.1 Vectors and $\mathbb{R}^n$-space

Section 1.1

1. Write a 3-column zero vector and a 5-column zero vector, respectively.

2. Write a 3-column nonzero vector and a 5-column nonzero vector, respectively.

3. Suppose that the buyer for a manufacturing plant must order different quantities of oil, paper, steel and plastics. He would order 40 units of oil, 50 units of paper, 80 units of steel and 20 units of plastics. Write the quantities in a single vector.

4. Suppose that a student's course marks for quiz 1, quiz 2, test 1, test 2 and final exam are 70, 85, 80, 75, 90, respectively. Write his marks as a column vector.

Solution. 1. $\begin{pmatrix} 0 \\ 0 \\ 0 \end{pmatrix}$ $\begin{pmatrix} 0 \\ 0 \\ 0 \\ 0 \\ 0 \end{pmatrix}$. 2. $\begin{pmatrix} 1 \\ 0 \\ -1 \end{pmatrix}$ $\begin{pmatrix} -1 \\ 0 \\ 1 \\ 0 \\ 1 \end{pmatrix}$.

3. $\begin{pmatrix} \text{oil} \\ \text{paper} \\ \text{steel} \\ \text{plastics} \end{pmatrix} = \begin{pmatrix} 40 \\ 50 \\ 80 \\ 20 \end{pmatrix}$. 4. $\begin{pmatrix} \text{quiz 1} \\ \text{quiz 2} \\ \text{test 1} \\ \text{test 2} \\ \text{final exam} \end{pmatrix} = \begin{pmatrix} 70 \\ 85 \\ 80 \\ 75 \\ 90 \end{pmatrix}$.

□

Section 1.2

1. Determine in which $\mathbb{R}^n$-space are the following vectors.

$$\overrightarrow{a} = \begin{pmatrix} 1 \\ 2 \\ 3 \end{pmatrix} \quad \overrightarrow{b} = \begin{pmatrix} 0 \\ 2 \\ 3 \\ 0 \end{pmatrix} \quad \overrightarrow{c} = \begin{pmatrix} 1 \\ 0 \\ 3 \\ 3 \\ 3 \\ 7 \end{pmatrix} \quad \overrightarrow{d} = \begin{pmatrix} 0 \\ 2 \\ 0 \\ 0 \\ 1 \end{pmatrix}$$

Solution. They are in $\mathbb{R}^3$, $\mathbb{R}^4$, $\mathbb{R}^6$, $\mathbb{R}^5$, respectively. □

Section 1.3

1. Let $\overrightarrow{a} = \begin{pmatrix} x-2y \\ 2x-y \\ 2z \end{pmatrix}$ and $\overrightarrow{b} = \begin{pmatrix} 2 \\ -2 \\ 1 \end{pmatrix}$. Find all $x, y, z \in \mathbb{R}$ such that $\overrightarrow{a} = \overrightarrow{b}$.

2. Let $\overrightarrow{a} = \begin{pmatrix} |x| \\ y^2 \end{pmatrix}$ and $\overrightarrow{b} = \begin{pmatrix} 1 \\ 4 \end{pmatrix}$. Find all $x, y \in \mathbb{R}$ such that $\overrightarrow{a} = \overrightarrow{b}$.

3. $\overrightarrow{a} = \begin{pmatrix} x-y \\ 4 \end{pmatrix}$ and $\overrightarrow{b} = \begin{pmatrix} 2 \\ x+y \end{pmatrix}$. Find all $x, y \in \mathbb{R}$ such that $\overrightarrow{a} - \overrightarrow{b}$ is a nonzero vector.

Solution. 1. $\overrightarrow{a} = \overrightarrow{b}$ if and only if

$$\begin{aligned} x-2y &= 2 - - - - - - - - (1) \\ 2x-y &= -2 - - - - - - - - (2) \\ 2z &= 1 - - - - - - - - - - (3) \end{aligned}$$

(1) $-$ (2) $\times$ 2: $(x-2y) - 2(2x-y) = 2 - 2(-2)$ and $x = -2$. By (2), we get $y = 2x + 2 = 2(-2) + 2 = -2$ and by (3), we have $z = \frac{1}{2}$ Therefore, when $x = -2$, $y = -2$ and $z = \frac{1}{2}$, $\overrightarrow{a} = \overrightarrow{b}$.

2. $\overrightarrow{a} = \overrightarrow{b}$ if and only if

$$\begin{cases} |x| = 1 & (1) \\ y^2 = 4 & (2) \end{cases}$$

By (1), we have $x = 1$ or $x = -1$ and by (2), we have $y = 2$ or $y = -2$. Hence, $\overrightarrow{a} = \overrightarrow{b}$ if $x = 1$ and $y = 2$; $x = 1$ and $y = -2$; $x = -1$ and $y = 2$; or $x = -1$ and $y = -2$.

3. We first find all x, y such that $\overrightarrow{a} - \overrightarrow{b} = \overrightarrow{0}$.

$$\overrightarrow{a}-\overrightarrow{b}=\begin{pmatrix}x-y\\4\end{pmatrix}-\begin{pmatrix}2\\x+y\end{pmatrix}=\begin{pmatrix}x-y-2\\4-x-y\end{pmatrix}$$

Let $\overrightarrow{a}-\overrightarrow{b}=\overrightarrow{0}$. Then $\begin{pmatrix}x-y-2\\4-x-y\end{pmatrix}=\begin{pmatrix}0\\0\end{pmatrix}$. Hence

$$\begin{cases}x-y-2=0 & (1)\\4-x-y=0 & (2)\end{cases}$$

$(1)+(2)$: $(x-y-2)+(4-x-y)=0$ and $y=1$. By (1), we get $x=y+2=1+2=3$. Hence when $x=2$ and $y=1$, $\overrightarrow{a}-\overrightarrow{b}=\overrightarrow{0}$. Now, we can find all x, y such that $\overrightarrow{a}-\overrightarrow{b}\neq\overrightarrow{0}$. When $x\neq 2$ or $y\neq 1$, $\overrightarrow{a}-\overrightarrow{b}\neq\overrightarrow{0}$, that is $\overrightarrow{a}-\overrightarrow{b}$ is a nonzero vector. □

Section 1.4

1. Let $\overrightarrow{a}=\begin{pmatrix}-1\\2\\3\end{pmatrix}$, $\overrightarrow{b}=\begin{pmatrix}2\\-2\\0\end{pmatrix}$, $\overrightarrow{c}=\begin{pmatrix}3\\0\\-1\end{pmatrix}$ and $\alpha\in\mathbb{R}$.

   Compute

   a. $\overrightarrow{a}+\overrightarrow{b}$ b. $0\overrightarrow{a}$ c. $-2\overrightarrow{b}$ d. $2\overrightarrow{a}-5\overrightarrow{b}$
   e. $2\overrightarrow{a}-\overrightarrow{b}+5\overrightarrow{c}$ f. $4\overrightarrow{a}+\alpha\overrightarrow{b}-2\overrightarrow{c}$.

2. Find x, y and z such that $\begin{pmatrix}9\\4y\\2z\end{pmatrix}+\begin{pmatrix}3x\\8\\-6\end{pmatrix}=\begin{pmatrix}0\\0\\0\end{pmatrix}$.

3. A company having 553 employees lists each employee's salary as a component of a vector $\overrightarrow{a}$ in $\mathbb{R}^{553}$. If a 6% salary increase has been approved, find the vector involving $\overrightarrow{a}$ that gives all the new salaries.

4. Let $\overrightarrow{a}=\begin{pmatrix}110\\88\\40\end{pmatrix}$ denote the current prices of three items at a store. Suppose that the store announces a sale so that the price of each item is reduced by 20%.

   (*a*) Find a 3-vector that gives the price changes for the three items.

   (*b*) Find a 3-vector that gives the new prices of the three items.

Solution.

1. a. $$\overrightarrow{a}+\overrightarrow{b}=\begin{pmatrix}-1\\2\\3\end{pmatrix}+\begin{pmatrix}2\\-2\\0\end{pmatrix}=\begin{pmatrix}-1+2\\2-2\\3+0\end{pmatrix}=\begin{pmatrix}1\\0\\3\end{pmatrix}.$$

b. $$0\overrightarrow{a}=0\begin{pmatrix}-1\\2\\3\end{pmatrix}=\begin{pmatrix}0(-1)\\0(2)\\0(3)\end{pmatrix}=\begin{pmatrix}0\\0\\0\end{pmatrix}.$$

c. $$-2\overrightarrow{b}=-2\begin{pmatrix}2\\-2\\0\end{pmatrix}=\begin{pmatrix}-2(2)\\-2(-2)\\-2(0)\end{pmatrix}=\begin{pmatrix}-4\\4\\0\end{pmatrix}.$$

d. $$2\overrightarrow{a}-5\overrightarrow{b}=2\begin{pmatrix}-1\\2\\3\end{pmatrix}-5\begin{pmatrix}2\\-2\\0\end{pmatrix}=\begin{pmatrix}-2\\4\\6\end{pmatrix}-\begin{pmatrix}10\\-10\\0\end{pmatrix}$$
$$=\begin{pmatrix}-2-10\\4-(-10)\\6-0\end{pmatrix}=\begin{pmatrix}-12\\14\\6\end{pmatrix}.$$

e. $$2\overrightarrow{a}-\overrightarrow{b}+5\overrightarrow{c}=2\begin{pmatrix}-1\\2\\3\end{pmatrix}-\begin{pmatrix}2\\-2\\0\end{pmatrix}+5\begin{pmatrix}3\\0\\-1\end{pmatrix}$$
$$=\begin{pmatrix}-2\\4\\6\end{pmatrix}-\begin{pmatrix}2\\-2\\0\end{pmatrix}+\begin{pmatrix}15\\0\\-5\end{pmatrix}=\begin{pmatrix}-2-2+15\\4+2+0\\6-0-5\end{pmatrix}$$
$$=\begin{pmatrix}11\\6\\1\end{pmatrix}.$$

f. $$4\overrightarrow{a}+\alpha\overrightarrow{b}-2\overrightarrow{c}=4\begin{pmatrix}-1\\2\\3\end{pmatrix}+\alpha\begin{pmatrix}2\\-2\\0\end{pmatrix}-2\begin{pmatrix}3\\0\\-1\end{pmatrix}$$
$$=\begin{pmatrix}-4\\8\\12\end{pmatrix}+\begin{pmatrix}2\alpha\\-2\alpha\\0\end{pmatrix}-\begin{pmatrix}6\\0\\-2\end{pmatrix}=\begin{pmatrix}-4+2\alpha-6\\8-2\alpha-0\\12+0+2\end{pmatrix}$$
$$=\begin{pmatrix}-10+2\alpha\\8-2\alpha\\14\end{pmatrix}.$$

2. Since $\begin{pmatrix}9+3x\\4y+8\\2z-6\end{pmatrix}=\begin{pmatrix}0\\0\\0\end{pmatrix}$, we have

$$\begin{cases} 9+3x=0 \\ 4y+8=0 \\ 2z-6=0 \end{cases}$$

This implies $x=-3$, $y=-2$, $z=3$.
3. $1.06\overrightarrow{a}$.

(4) (a) $-0.2\overrightarrow{a} = \begin{pmatrix} -22 \\ -17.6 \\ -8 \end{pmatrix}$. (b) $0.8\overrightarrow{a} = \begin{pmatrix} 88 \\ 70.4 \\ 32 \end{pmatrix}$. □

Section 1.5

1. Let $\overrightarrow{a_1} = \begin{pmatrix} 1 \\ 2 \end{pmatrix}$, $\overrightarrow{a_2} = \begin{pmatrix} 1 \\ 1 \end{pmatrix}$ and $\overrightarrow{b} = \begin{pmatrix} 1 \\ 0 \end{pmatrix}$. Show whether $\overrightarrow{b}$ is a linear combination of $\overrightarrow{a_1}$ and $\overrightarrow{a_2}$.

2. Let $\overrightarrow{a_1} = \begin{pmatrix} 1 \\ 0 \\ 1 \end{pmatrix}$, $\overrightarrow{a_2} = \begin{pmatrix} 1 \\ 3 \\ 2 \end{pmatrix}$ and $\overrightarrow{a_3} = \begin{pmatrix} -1 \\ 1 \\ 3 \end{pmatrix}$.

(i) Let $\overrightarrow{b} = \begin{pmatrix} -1 \\ -9 \\ -4 \end{pmatrix}$. Verify whether $2\overrightarrow{a_1} - 3\overrightarrow{a_2} = \overrightarrow{b}$.

(ii) Let $\overrightarrow{b} = \begin{pmatrix} 3 \\ 6 \\ 6 \end{pmatrix}$. Verify whether $\overrightarrow{a_1} - 2\overrightarrow{a_2} = \overrightarrow{b}$.

(iii) Let $\overrightarrow{b} = \begin{pmatrix} 2 \\ 7 \\ 7 \end{pmatrix}$. Verify whether $\overrightarrow{a_1} + 2\overrightarrow{a_2} + \overrightarrow{a_3} = \overrightarrow{b}$.

3. Let $\overrightarrow{a_1} = \begin{pmatrix} 1 \\ 1 \end{pmatrix}$ and $\overrightarrow{a_2} = \begin{pmatrix} 2 \\ 1 \end{pmatrix}$. Show whether $\overrightarrow{b} = \begin{pmatrix} 0 \\ 1 \end{pmatrix}$ is a linear combination of $\overrightarrow{a_1}$ and $\overrightarrow{a_2}$.

Solution. 1. Let $x\overrightarrow{a_1} + y\overrightarrow{a_2} = \overrightarrow{b}$. Then

$$x\begin{pmatrix} 1 \\ 2 \end{pmatrix} + y\begin{pmatrix} 1 \\ 1 \end{pmatrix} = \begin{pmatrix} 1 \\ 0 \end{pmatrix}$$

$$\begin{pmatrix} x \\ 2x \end{pmatrix} + \begin{pmatrix} y \\ y \end{pmatrix} = \begin{pmatrix} x+y \\ 2x+y \end{pmatrix} = \begin{pmatrix} 1 \\ 0 \end{pmatrix}$$

Hence,

$$\begin{cases} x+y=1 & (1) \\ 2x+y=0 & (2) \end{cases}$$

(1) − (2): $(x+y)-(2x+y)=1-0$ and $x=-1$. By (1),

$$y=1-x=1-(-1)=1+1=2.$$

Hence, $-\overrightarrow{a_1}+2\overrightarrow{a_2}=\overrightarrow{b}$ and $\overrightarrow{b}$ is a linear combination of $\overrightarrow{a_1}$ and $\overrightarrow{a_2}$.

2. (i) $$2\overrightarrow{a_1}-3\overrightarrow{a_2}=2\begin{pmatrix}1\\0\\1\end{pmatrix}-3\begin{pmatrix}1\\3\\2\end{pmatrix}=\begin{pmatrix}2\\0\\2\end{pmatrix}-\begin{pmatrix}3\\9\\6\end{pmatrix}=\begin{pmatrix}2-3\\0-9\\2-6\end{pmatrix}$$
$$=\begin{pmatrix}-1\\-9\\-4\end{pmatrix}=\overrightarrow{b}.$$

(ii) $$\overrightarrow{a_1}-2\overrightarrow{a_2}=\begin{pmatrix}1\\0\\1\end{pmatrix}-2\begin{pmatrix}1\\3\\2\end{pmatrix}=\begin{pmatrix}1-2\\0-6\\1-4\end{pmatrix}=\begin{pmatrix}-1\\-6\\-3\end{pmatrix}\neq\overrightarrow{b}.$$

(iii) $$\overrightarrow{a_1}+2\overrightarrow{a_2}+\overrightarrow{a_3}=\begin{pmatrix}1\\0\\1\end{pmatrix}+2\begin{pmatrix}1\\3\\2\end{pmatrix}+\begin{pmatrix}-1\\1\\3\end{pmatrix}=\begin{pmatrix}1+2-1\\0+6+1\\1+4+3\end{pmatrix}$$
$$=\begin{pmatrix}2\\7\\8\end{pmatrix}\neq\overrightarrow{b}.$$

3. Let $x\overrightarrow{a_1}+y\overrightarrow{a_2}=\overrightarrow{b}$. Then

$$x\begin{pmatrix}1\\1\end{pmatrix}+y\begin{pmatrix}2\\1\end{pmatrix}=\begin{pmatrix}0\\1\end{pmatrix}.$$

Hence,

$$\begin{cases}x+2y=0 & (1)\\ x+y=1 & (2)\end{cases}$$

(1) − (2) implies that $(x+2y)-(x+y)=0-1$ and $y=-1$. By (1), $x=-2y=-2(-1)=2$. Hence, $2\overrightarrow{a_1}-\overrightarrow{a_2}=\overrightarrow{b}$ and $\overrightarrow{b}$ is a linear combination of $\overrightarrow{a_1}$ and $\overrightarrow{a_2}$. □

Section 1.6

1. Let $\overrightarrow{a}=\begin{pmatrix}-1\\2\\3\end{pmatrix}$, $\overrightarrow{b}=\begin{pmatrix}2\\-2\\0\end{pmatrix}$, $\overrightarrow{c}=\begin{pmatrix}3\\0\\-1\end{pmatrix}$ and $\alpha\in\mathbb{R}$.

   Compute

   (a). $\overrightarrow{a}\cdot\overrightarrow{b}$; (b). $\overrightarrow{a}\cdot\overrightarrow{c}$; (c). $\overrightarrow{b}\cdot\overrightarrow{c}$; (d). $\overrightarrow{a}\cdot(\overrightarrow{b}+\overrightarrow{c})$.

2. Let $\overrightarrow{u} = (1, -1, 0, 2)$, $\overrightarrow{v} = (-2, -1, 1, -4)$ and $\overrightarrow{w} = (3, 2, -1, 0)$.

   (*a*) Find a vector $\overrightarrow{a} \in \mathbb{R}^4$ such that $2\overrightarrow{u} - 3\overrightarrow{v} - \overrightarrow{a} = \overrightarrow{w}$.

   (*b*) Compute $\overrightarrow{u} \cdot \overrightarrow{v}$, $\overrightarrow{u} \cdot \overrightarrow{w}$ and $\overrightarrow{w} \cdot \overrightarrow{v}$.

   (*c*) Find a vector $\overrightarrow{a}$ such that

$$\frac{1}{2}[2\overrightarrow{u} - 3\overrightarrow{v} + \overrightarrow{a}] = 2\overrightarrow{w} + \overrightarrow{u} - 2\overrightarrow{a}.$$

3. Assume that the percentages for homework, test 1, test 2 and final exam for a course are 10%, 25%, 25%, 40%, respectively. The total marks for homework, test 1, test 2 and final exam are 10, 50, 50, 90, respectively. A student's corresponding marks are 8, 46, 48, 81, respectively. What are the student's final marks out of 100.

Solution.

1. (*a*). $\overrightarrow{a} \cdot \overrightarrow{b} = \begin{pmatrix} -1 & 2 & 3 \end{pmatrix} \begin{pmatrix} 2 \\ -2 \\ 0 \end{pmatrix} = (-1)(2) + 2(-2) + (3)(0) = -6.$

   (*b*). $\overrightarrow{a} \cdot \overrightarrow{c} = \begin{pmatrix} -1 & 2 & 3 \end{pmatrix} \begin{pmatrix} 3 \\ 0 \\ -1 \end{pmatrix} = (-1)(3) + 2(0) + 3(-1) = -6.$

   (*c*). $\overrightarrow{b} \cdot \overrightarrow{c} = \begin{pmatrix} 2 & -2 & 0 \end{pmatrix} \begin{pmatrix} 3 \\ 0 \\ -1 \end{pmatrix} = (2)(3) + (-2)(0) + (0)(-1) = 6.$

   (*d*). $\overrightarrow{a} \cdot (\overrightarrow{b} + \overrightarrow{c}) = \begin{pmatrix} -1 & 2 & 3 \end{pmatrix} \left(\begin{pmatrix} 2 \\ -2 \\ 0 \end{pmatrix} + \begin{pmatrix} 3 \\ 0 \\ -1 \end{pmatrix} \right)$

$$= \begin{pmatrix} -1 & 2 & 3 \end{pmatrix} \begin{pmatrix} 5 \\ -2 \\ -1 \end{pmatrix} = (-1)(5) + 2(-2) + 3(-1) = -12.$$

2. (*a*) Since $2\overrightarrow{u} - 3\overrightarrow{v} - \overrightarrow{a} = \overrightarrow{w}$, we have

$$\begin{aligned} \overrightarrow{a} &= 2\overrightarrow{u} - 3\overrightarrow{v} - \overrightarrow{w} \\ &= 2(1, -1, 0, 2) - 3(-2, -1, 1, -4) - (3, 2, -1, 0) \\ &= (2, -2, 0, 4) - (-6, -3, 3, -12) - (3, 2, -1, 0) \\ &= (2 + 6 - 3, -2 + 3 - 2, 0 - 3 + 1, 4 + 12 - 0) = (5, -1, -2, 16). \end{aligned}$$

$$(b)\quad \overrightarrow{u}\cdot\overrightarrow{v} = \begin{pmatrix} 1 & -1 & 0 & 2 \end{pmatrix}\begin{pmatrix} -2 \\ -1 \\ 1 \\ -4 \end{pmatrix} = 1(-2)+(-1)(-1)+0(1)+2(-4)$$

$$= -2+1+0-8 = -9.$$

$$\overrightarrow{u}\cdot\overrightarrow{w} = \begin{pmatrix} 1 & -1 & 0 & 2 \end{pmatrix}\begin{pmatrix} 3 \\ 2 \\ -1 \\ 0 \end{pmatrix} = 1(3)+(-1)2+0(-1)+2(0)$$

$$= 3-2+0+0 = 1.$$

$$\overrightarrow{w}\cdot\overrightarrow{v} = \begin{pmatrix} 3 & 2 & -1 & 0 \end{pmatrix}\begin{pmatrix} -2 \\ -1 \\ 1 \\ -4 \end{pmatrix} = 3(-2)+2(-1)+(-1)(1)+0(-4)$$

$$= -6-2-1+0 = -9.$$

(c) Since $\frac{1}{2}[2\overrightarrow{u} - 3\overrightarrow{v} + \overrightarrow{a}] = 2\overrightarrow{w} + \overrightarrow{u} - 2\overrightarrow{a}$, we have

$$\begin{aligned} \overrightarrow{u} - \frac{3}{2}\overrightarrow{v} + \frac{1}{2}\overrightarrow{a} &= 2\overrightarrow{w} + \overrightarrow{u} - 2\overrightarrow{a} \\ 2\overrightarrow{a} + \frac{1}{2}\overrightarrow{a} &= 2\overrightarrow{w} + \overrightarrow{u} - \overrightarrow{u} + \frac{3}{2}\overrightarrow{v} = 2\overrightarrow{w} + \frac{3}{2}\overrightarrow{v} \\ \frac{5}{2}\overrightarrow{a} &= 2\overrightarrow{w} + \frac{3}{2}\overrightarrow{v}. \end{aligned}$$

This implies that

$$\begin{aligned} \overrightarrow{a} &= \frac{2}{5}(2\overrightarrow{w} + \frac{3}{2}\overrightarrow{v}) = \frac{1}{5}[4\overrightarrow{w} + 3\overrightarrow{v}] \\ &= \frac{1}{5}\left[4\begin{pmatrix} 3 & 2 & -1 & 0 \end{pmatrix} + 3\begin{pmatrix} -2 & -1 & 1 & -4 \end{pmatrix}\right] \\ &= \frac{1}{5}\left[\begin{pmatrix} 12 & 8 & -4 & 0 \end{pmatrix} + \begin{pmatrix} -6 & -3 & 3 & -12 \end{pmatrix}\right] \\ &= \frac{1}{5}\begin{pmatrix} 12-6 & 8-3 & -4+3 & 0-12 \end{pmatrix} \\ &= \frac{1}{5}\begin{pmatrix} 6 & 5 & -1 & -12 \end{pmatrix} = \begin{pmatrix} \frac{6}{5} & 1 & -\frac{1}{5} & -\frac{12}{5} \end{pmatrix}. \end{aligned}$$

3. Let $\overrightarrow{a} = 100(8/10, 46/50, 48/50, 81/90)$ and $\overrightarrow{b} = (10\%, 25\%, 25\%, 40\%)$.

Then the total final marks out of 100 for the student is

$$\begin{aligned}\vec{a}\cdot\vec{b} &= 100(8/10, 46/50, 48/50, 81/90)\cdot(10\%, 25\%, 25\%, 40\%)\\ &= (8/10, 46/50, 48/50, 81/90)\cdot(10, 25, 25, 40)\\ &= 8+23+24+36 = 91.\end{aligned}$$

□

A.2 Matrices

Section 2.1

1. Find the sizes of the following matrices:

   $(i)\ A=(2);\quad (ii)\ B=\begin{pmatrix}5 & 8\\ 1 & 4\end{pmatrix};\quad (iii)\ C=\begin{pmatrix}10 & 3 & 8\\ 10 & 3 & 4\end{pmatrix};$

   $(iv)\ D=\begin{pmatrix}1 & 0 & 0\\ 0 & 10 & 2\\ 0 & 0 & 1\\ 6 & 9 & 10\end{pmatrix}.$

2. Use column vectors and row vectors to rewrite each of the following matrices.

$$A=\begin{pmatrix}3 & 3 & 2\\ 1 & 8 & 1\\ 5 & 4 & 10\end{pmatrix};\ B=\begin{pmatrix}1 & 9 & 7 & 2\\ 3 & 10 & 8 & 7\\ 4 & 3 & 10 & 4\end{pmatrix};\ C=\begin{pmatrix}9 & 8 & 7\\ 6 & 5 & 4\\ 3 & 2 & 1\\ 0 & -1 & -2\end{pmatrix}.$$

Solution.

1. (i) 1×1; (ii) 2×2; (iii) 2×3; (iv) 4×3

2. Let $\vec{c_1}=\begin{pmatrix}3\\1\\5\end{pmatrix}$, $\vec{c_2}=\begin{pmatrix}3\\8\\4\end{pmatrix}$, $\vec{c_3}=\begin{pmatrix}2\\1\\10\end{pmatrix}$. Then

$$A=\begin{pmatrix}\vec{c_1} & \vec{c_2} & \vec{c_3}\end{pmatrix}.$$

Let $\vec{r_1}=\begin{pmatrix}3 & 3 & 2\end{pmatrix}$, $\vec{r_2}=\begin{pmatrix}1 & 8 & 1\end{pmatrix}$, $\vec{r_3}=\begin{pmatrix}5 & 4 & 10\end{pmatrix}$. Then

$$A=\begin{pmatrix}\vec{r_1}\\ \vec{r_2}\\ \vec{r_3}\end{pmatrix}.$$

Let $\overrightarrow{c_1} = \begin{pmatrix} 1 \\ 3 \\ 4 \end{pmatrix}$, $\overrightarrow{c_2} = \begin{pmatrix} 9 \\ 10 \\ 3 \end{pmatrix}$, $\overrightarrow{c_3} = \begin{pmatrix} 7 \\ 8 \\ 10 \end{pmatrix}$, $\overrightarrow{c_4} = \begin{pmatrix} 2 \\ 7 \\ 4 \end{pmatrix}$. Then

$$B = \begin{pmatrix} \overrightarrow{c_1} & \overrightarrow{c_2} & \overrightarrow{c_3} & \overrightarrow{c_4} \end{pmatrix}.$$

Let $\overrightarrow{r_1} = \begin{pmatrix} 1 & 9 & 7 & 2 \end{pmatrix}$, $\overrightarrow{r_2} = \begin{pmatrix} 3 & 10 & 8 & 7 \end{pmatrix}$, $\overrightarrow{r_3} = \begin{pmatrix} 4 & 3 & 10 & 4 \end{pmatrix}$. Then

$$B = \begin{pmatrix} \overrightarrow{r_1} \\ \overrightarrow{r_2} \\ \overrightarrow{r_3} \end{pmatrix}.$$

Let $\overrightarrow{c_1} = \begin{pmatrix} 9 \\ 6 \\ 3 \\ 0 \end{pmatrix}$, $\overrightarrow{c_2} = \begin{pmatrix} 8 \\ 5 \\ 2 \\ -1 \end{pmatrix}$ and $\overrightarrow{c_3} = \begin{pmatrix} 7 \\ 4 \\ 1 \\ -2 \end{pmatrix}$. Then

$$C = \begin{pmatrix} \overrightarrow{c_1} & \overrightarrow{c_2} & \overrightarrow{c_3} \end{pmatrix}.$$

Let $\overrightarrow{r_1} = \begin{pmatrix} 9 & 8 & 7 \end{pmatrix}$, $\overrightarrow{r_2} = \begin{pmatrix} 6 & 5 & 4 \end{pmatrix}$, $\overrightarrow{r_3} = \begin{pmatrix} 3 & 2 & 1 \end{pmatrix}$ and $\overrightarrow{r_4} = \begin{pmatrix} 0 & -1 & -2 \end{pmatrix}$. Then

$$C = \begin{pmatrix} \overrightarrow{r_1} \\ \overrightarrow{r_2} \\ \overrightarrow{r_3} \\ \overrightarrow{r_4} \end{pmatrix}.$$

□

Section 2.2

1. Let

$$A_1 = \begin{pmatrix} 0 & -1 \\ -1 & 0 \end{pmatrix}, A_2 = \begin{pmatrix} 1 & 4 & 5 \\ 4 & -3 & -1 \\ 5 & 0 & 7 \end{pmatrix}, A_3 = \begin{pmatrix} y & x^3 & 1 \\ x^3 & y & x \\ 1 & x & z \end{pmatrix}.$$

   (i) Find the trace of each of the above matrices.

   (ii) Determine which of the above matrices are symmetric.

2. Identify whether the given matrix is a lower triangular, upper triangular, diagonal, triangular or identity matrix.

$$A_1 = \begin{pmatrix} 0 & 2 \\ 0 & 0 \end{pmatrix} \quad A_2 = \begin{pmatrix} 1 & 0 & 0 \\ 2 & 0 & 0 \\ 0 & 0 & 1 \end{pmatrix} \quad A_3 = \begin{pmatrix} 0 & 0 & 0 \\ 0 & 0 & 0 \\ 0 & 0 & 0 \end{pmatrix}$$

$$A_4 = \begin{pmatrix} 1 & 0 & x \\ 1 & 0 & 0 \\ x & 0 & 0 \end{pmatrix} \quad A_5 = \begin{pmatrix} 0 & 1 \\ 1 & 0 \end{pmatrix} \quad A_6 = \begin{pmatrix} 1 & 0 & 1 \\ 0 & 0 & 0 \\ 0 & 0 & 1 \end{pmatrix}$$

$$A_7 = \begin{pmatrix} 2 & 0 \\ 0 & 2 \end{pmatrix} \quad A_8 = \begin{pmatrix} 1 & 0 & 0 \\ 0 & 0 & 0 \\ 2 & 0 & 1 \end{pmatrix} \quad A_9 = \begin{pmatrix} 1 & 0 & 0 \\ 0 & 2 & 0 \\ -1 & 1 & 1 \end{pmatrix}$$

$$A_{10} = \begin{pmatrix} 1 & 0 & 1 \\ 2 & 0 & 0 \\ 1 & 0 & 1 \end{pmatrix} \quad A_{11} = \begin{pmatrix} 0 & 0 \\ 0 & 0 \\ 0 & 0 \end{pmatrix} \quad A_{12} = \begin{pmatrix} 0 & 0 & 1 \\ 0 & 1 & 0 \\ 1 & 0 & 0 \end{pmatrix}$$

Solution.

1. (*i*) $\mathrm{tr}(A_1) = 0 + 0 = 0$, $\mathrm{tr}(A_2) = 1 + (-3) + 7 = 5$ and

$$\mathrm{tr}(A_3) = y + y + z = 2y + z.$$

(*ii*) A_1 is symmetric, A_2 is not symmetric and A_3 is symmetric.

2. lower triangular: A_2, A_3, A_7, A_8, A_9;
upper triangular: A_1, A_3, A_6, A_7;
diagonal: A_3, A_7;
triangular: $A_1, A_2, A_3, A_6, A_7, A_8, A_9$.
There are no identity matrices. □

Section 2.3

1. Which of the following matrices are row echelon matrices?

$$A = \begin{pmatrix} 1 & 0 \\ 0 & 1 \end{pmatrix} \quad B = \begin{pmatrix} 9 & 4 & 3 & 5 & 9 \\ 0 & 0 & 0 & 7 & 4 \end{pmatrix} \quad C = \begin{pmatrix} 2 & 5 & 0 \\ 0 & 9 & 0 \\ 0 & 0 & 8 \end{pmatrix}$$

$$D = \begin{pmatrix} 3 & 0 & 1 & 6 & 0 \\ 0 & 0 & 8 & 8 & 6 \\ 0 & 0 & 0 & 0 & 1 \\ 0 & 0 & 0 & 0 & 0 \end{pmatrix} \quad E = \begin{pmatrix} 1 & 0 & 0 \\ 0 & 0 & 1 \\ 0 & 1 & 0 \end{pmatrix} \quad F = \begin{pmatrix} 1 & 4 & 1 \\ 0 & 0 & 0 \\ 0 & 6 & 6 \end{pmatrix}$$

$$G = \begin{pmatrix} 0 & 8 & 1 & 7 \\ 2 & 9 & 8 & 0 \\ 0 & 0 & 0 & 0 \end{pmatrix} \quad H = \begin{pmatrix} 0 & 0 & 0 & 0 \\ 4 & 0 & 5 & 8 \\ 0 & 9 & 6 & 1 \end{pmatrix}$$

2. Which of the following matrices are reduced row echelon matrices?

$$A = \begin{pmatrix} 1 & 0 & 0 & 1 \\ 0 & 1 & 0 & 1 \\ 0 & 0 & 1 & 0 \\ 0 & 0 & 0 & 0 \end{pmatrix} \quad B = \begin{pmatrix} 1 & 5 & 0 & 6 & 0 \\ 0 & 0 & 1 & 3 & 0 \\ 0 & 0 & 0 & 0 & 1 \\ 0 & 0 & 0 & 0 & 0 \end{pmatrix}$$

$$C = \begin{pmatrix} 1 & 1 & 1 & 1 & 1 \\ 0 & 0 & 0 & 1 & 1 \end{pmatrix} \quad D = \begin{pmatrix} 2 & 0 & 1 \\ 0 & 1 & 0 \\ 0 & 0 & 1 \end{pmatrix}$$

$$E = \begin{pmatrix} 1 & 0 & 0 & 4 & 0 \\ 0 & 1 & 1 & 3 & 0 \\ 0 & 0 & 0 & 0 & 1 \\ 0 & 0 & 0 & 0 & 0 \end{pmatrix} \quad F = \begin{pmatrix} 1 & 0 & 0 & 2 & 0 \\ 0 & 0 & 2 & 1 & 0 \end{pmatrix}$$

Solution.

1. A, B, C, D are row echelon matrices.
2. A, B, E are reduced row echelon matrices. □

Section 2.4

1. Find the transposes of the following matrices:

$$A = \begin{pmatrix} 33 \\ 8 \\ 12 \end{pmatrix} \quad B = \begin{pmatrix} 3 & 11 & 2 \end{pmatrix} \quad C = (4) \quad D = \begin{pmatrix} 3 & 9 & -19 \\ -2 & 8 & -7 \\ 5 & 3 & -9 \end{pmatrix}$$

Solution. 1. $A^T = \begin{pmatrix} 33 & 8 & 12 \end{pmatrix}$, $B^T = \begin{pmatrix} 3 \\ 11 \\ 2 \end{pmatrix}$, $C^T = (4)$ and

$$D^T = \begin{pmatrix} 3 & -2 & 5 \\ 9 & 8 & 3 \\ -19 & -7 & -9 \end{pmatrix}.$$

□

Section 2.5

1. Let $A = \begin{pmatrix} 9 & 3 \\ 6 & x^2 \end{pmatrix}$ and $B = \begin{pmatrix} 9 & 3 \\ 6 & 4 \end{pmatrix}$. Find all $x \in \mathbb{R}$ such that $A = B$.

2. Let $C = \begin{pmatrix} 12 & 5 & 25 \\ 19 & 4 & 6 \end{pmatrix}$ and $D = \begin{pmatrix} 12 & 5 & x^2 \\ 19 & 4 & 6 \end{pmatrix}$. Find all $x \in \mathbb{R}$ such that $C = D$.

3. Let $E = \begin{pmatrix} 120 & 25 & 122 \\ 123 & 124 & 125 \\ 126 & 127 & 128 \end{pmatrix}$ and $F = \begin{pmatrix} 120 & x^2 & 122 \\ 123 & 124 & x^3 \\ 26x - 4 & 127 & 128 \end{pmatrix}$. Find all $x \in \mathbb{R}$ such that $E = F$.

4. Let $A = \begin{pmatrix} a & b \\ 3x + 2y & -x + y \end{pmatrix}$ and $B = \begin{pmatrix} 1 & 1 \\ 3 & 6 \end{pmatrix}$. Find all $a, b, x, y \in \mathbb{R}$ such that $A = B$.

5. Let

$$C = \begin{pmatrix} a + b & 2b - a \\ x - 2y & 5x + 3y \end{pmatrix} \quad \text{and } D = \begin{pmatrix} 6 & 0 \\ 8 & 14 \end{pmatrix}.$$

Find all $a, b, x, y \in \mathbb{R}$ such that $C = D$.

Solution. 1. Let $A = B$. Then $x^2 = 4$. This implies $x = 2$ or $x = -2$.

2. Let $C = D$. Then $x^2 = 25$. Solving the equation, we obtain $x = 5$ or $x = -5$.

3. Let $E = F$. Then

$$\begin{cases} x^2 = 25 & (1) \\ x^3 = 125 & (2) \\ 26x - 4 = 126 & (3) \end{cases}$$

By (1), $x = 5$ or $x = -5$. By (2), $x = 5$. By (3), $26x = 126 + 4 = 130$ and $x = \frac{130}{26} = 5$. Hence, when $x = 5$, $E = F$.

4. Let $A = B$. Then $a = 1$ and $b = 1$. Hence,

$$\begin{cases} 3x + 2y = 3 & (1) \\ -x + y = 6 & (2) \end{cases} \tag{A.1}$$

$(1) - (2) \times 2$ implies that $3x + 2y - 2(-x + y) = 3 - 12$ and $5x = -9$. Hence, $x = -\frac{9}{5}$. By (2), we obtain

$$y = x + 6 = -\frac{9}{5} + 6 = \frac{-9 + 30}{5} = \frac{21}{5}.$$

Hence when $a = 1, b = 1, x = -\frac{9}{5}, y = \frac{21}{5}, A = B$.

5. Let $C = D$. Then

$$\begin{cases} a + b = 6 & (1) \\ 2b - a = 0 & (2) \\ x - 2y = 8 & (3) \\ 5x + 3y = 14. & (4) \end{cases}$$

$(1) + (2)$ implies that $(a + b) + (2b - a) = 6 + 0 = 6$ and $b = 2$. By (1), we have $a = 6 - b = 6 - 2 = 4$.

$(3) \times 5 - (4)$ implies that $5(x - 2y) - (5x + 3y) = 40 - 14$ and $-13y = 26$. Hence, $y = -2$. By (3), we get $x = 8 + 2y = 8 + 2(-2) = 8 - 4 = 4$. Hence, when $a = 4, b = 2, x = 4$, and $y = -2, C = D$. □

Section 2.6

1. Let $A = \begin{pmatrix} -2 & 3 & 4 \\ 6 & -1 & -8 \end{pmatrix}$ and $B = \begin{pmatrix} 7 & -8 & 9 \\ 0 & -1 & 0 \end{pmatrix}$. Compute

(*i*) $A + B$; (*ii*) $-A$; (*iii*) $4A - 2B$; (*iv*) $100A + B$.

2. Let $A = \begin{pmatrix} 9 & 5 & 1 \\ 8 & 0 & 0 \\ 0 & 3 & 2 \end{pmatrix}$, $B = \begin{pmatrix} 2 & 2 & 2 \\ 0 & 0 & 0 \\ 4 & 6 & 8 \end{pmatrix}$ and $C = \begin{pmatrix} 4 & 7 & 0 \\ 0 & 3 & 1 \\ 0 & 0 & 2 \end{pmatrix}$.

Compute (1) $3A - 2B + C$; (2) $[3(A + B)]^T$; (3) $(4A + \frac{1}{2}B - 3C)^T$.

3. Find the matrix A if $\left[(3A^T) - \begin{pmatrix} -7 & -2 \\ -6 & 9 \end{pmatrix}^T\right]^T = \begin{pmatrix} -5 & -10 \\ 33 & 12 \end{pmatrix}$.

4. Find the matrix B if

$$\left[\frac{1}{2}B + \begin{pmatrix} 6 & 3 \\ 8 & 3 \\ 1 & 4 \end{pmatrix}\right]^T - 3\begin{pmatrix} -5 & 6 \\ 8 & -9 \\ -4 & 2 \end{pmatrix}^T = \begin{pmatrix} 23 & -16 & 17 \\ -16 & 26.5 & 2 \end{pmatrix}.$$

Solution.

1. (*i*)
$$\begin{aligned} A + B &= \begin{pmatrix} -2 & 3 & 4 \\ 6 & -1 & -8 \end{pmatrix} + \begin{pmatrix} 7 & -8 & 9 \\ 0 & -1 & 0 \end{pmatrix} \\ &= \begin{pmatrix} -2+7 & 3-8 & 4+9 \\ 6+0 & -1-1 & -8+0 \end{pmatrix} = \begin{pmatrix} 5 & -5 & 13 \\ 6 & -2 & -8 \end{pmatrix}. \end{aligned}$$

(*ii*)
$$-A = -\begin{pmatrix} -2 & 3 & 4 \\ 6 & -1 & -8 \end{pmatrix} = \begin{pmatrix} 2 & -3 & -4 \\ -6 & 1 & 8 \end{pmatrix}.$$

(*iii*)
$$\begin{aligned} 4A - 2B &= 4\begin{pmatrix} -2 & 3 & 4 \\ 6 & -1 & -8 \end{pmatrix} - 2\begin{pmatrix} 7 & -8 & 9 \\ 0 & -1 & 0 \end{pmatrix} \\ &= \begin{pmatrix} -8 & 12 & 16 \\ 24 & -4 & -32 \end{pmatrix} - \begin{pmatrix} 14 & -16 & 18 \\ 0 & -2 & 0 \end{pmatrix} \\ &= \begin{pmatrix} -8-14 & 12+16 & 16-18 \\ 24-0 & -4+2 & -32-0 \end{pmatrix} \\ &= \begin{pmatrix} -22 & 28 & -2 \\ 24 & -2 & -32 \end{pmatrix}. \end{aligned}$$

(*iv*)
$$\begin{aligned} 100A + B &= 100\begin{pmatrix} -2 & 3 & 4 \\ 6 & -1 & -8 \end{pmatrix} + \begin{pmatrix} 7 & -8 & 9 \\ 0 & -1 & 0 \end{pmatrix} \\ &= \begin{pmatrix} -200 & 300 & 400 \\ 600 & -100 & -800 \end{pmatrix} + \begin{pmatrix} 7 & -8 & 9 \\ 0 & -1 & 0 \end{pmatrix} \\ &= \begin{pmatrix} -200+7 & 300-8 & 400+9 \\ 600+0 & -100-1 & -800+0 \end{pmatrix} \\ &= \begin{pmatrix} -193 & 292 & 409 \\ 600 & -101 & -800 \end{pmatrix}. \end{aligned}$$

2. (1)

$$
\begin{aligned}
3A - 2B + C &= 3\begin{pmatrix} 9 & 5 & 1 \\ 8 & 0 & 0 \\ 0 & 3 & 2 \end{pmatrix} - 2\begin{pmatrix} 2 & 2 & 2 \\ 0 & 0 & 0 \\ 4 & 6 & 8 \end{pmatrix} + \begin{pmatrix} 4 & 7 & 0 \\ 0 & 3 & 1 \\ 0 & 0 & 2 \end{pmatrix} \\
&= \begin{pmatrix} 27 & 15 & 3 \\ 24 & 0 & 0 \\ 0 & 9 & 6 \end{pmatrix} - \begin{pmatrix} 4 & 4 & 4 \\ 0 & 0 & 0 \\ 8 & 12 & 16 \end{pmatrix} + \begin{pmatrix} 4 & 7 & 0 \\ 0 & 3 & 1 \\ 0 & 0 & 2 \end{pmatrix} \\
&= \begin{pmatrix} 27-4 & 15-4 & 3-4 \\ 24-0 & 0-0 & 0-0 \\ 0-8 & 9-12 & 6-16 \end{pmatrix} + \begin{pmatrix} 4 & 7 & 0 \\ 0 & 3 & 1 \\ 0 & 0 & 2 \end{pmatrix} \\
&= \begin{pmatrix} 23 & 11 & -1 \\ 24 & 0 & 0 \\ -8 & -3 & -10 \end{pmatrix} + \begin{pmatrix} 4 & 7 & 0 \\ 0 & 3 & 1 \\ 0 & 0 & 2 \end{pmatrix} \\
&= \begin{pmatrix} 23+4 & 11+7 & -1+0 \\ 24+0 & 0+3 & 0+1 \\ -8+0 & -3+0 & -10+2 \end{pmatrix} = \begin{pmatrix} 27 & 18 & -1 \\ 24 & 3 & 1 \\ -8 & -3 & -8 \end{pmatrix}.
\end{aligned}
$$

(2)

$$
\begin{aligned}
[3(A+B)]^T &= 3(A+B)^T = 3(A^T + B^T) \\
&= 3\left[\begin{pmatrix} 9 & 8 & 0 \\ 5 & 0 & 3 \\ 1 & 0 & 2 \end{pmatrix} + \begin{pmatrix} 2 & 0 & 4 \\ 2 & 0 & 6 \\ 2 & 0 & 8 \end{pmatrix}\right] \\
&= 3\begin{pmatrix} 11 & 8 & 4 \\ 7 & 0 & 9 \\ 3 & 0 & 10 \end{pmatrix} = \begin{pmatrix} 33 & 24 & 12 \\ 21 & 0 & 27 \\ 9 & 0 & 30 \end{pmatrix}.
\end{aligned}
$$

(3)

$$
\begin{aligned}
(4A + \frac{1}{2}B - 3C)^T &= 4A^T + \frac{1}{2}B^T - 3C^T \\
&= 4\begin{pmatrix} 9 & 8 & 0 \\ 5 & 0 & 3 \\ 1 & 0 & 2 \end{pmatrix} + \frac{1}{2}\begin{pmatrix} 2 & 0 & 4 \\ 2 & 0 & 6 \\ 2 & 0 & 8 \end{pmatrix} - 3\begin{pmatrix} 4 & 0 & 0 \\ 7 & 3 & 0 \\ 0 & 1 & 2 \end{pmatrix} \\
&= \begin{pmatrix} 36 & 32 & 0 \\ 20 & 0 & 12 \\ 4 & 0 & 8 \end{pmatrix} + \begin{pmatrix} 1 & 0 & 2 \\ 1 & 0 & 3 \\ 1 & 0 & 4 \end{pmatrix} - \begin{pmatrix} \frac{4}{3} & 0 & 0 \\ \frac{7}{3} & 1 & 0 \\ 0 & \frac{1}{3} & \frac{2}{3} \end{pmatrix} \\
&= \begin{pmatrix} 36+1 & 32+0 & 0+2 \\ 20+1 & 0+0 & 12+3 \\ 4+1 & 0+0 & 8+4 \end{pmatrix} - \begin{pmatrix} \frac{4}{3} & 0 & 0 \\ \frac{7}{3} & 1 & 0 \\ 0 & \frac{1}{3} & \frac{2}{3} \end{pmatrix} \\
&= \begin{pmatrix} 37 & 32 & 2 \\ 21 & 0 & 15 \\ 5 & 0 & 12 \end{pmatrix} - \begin{pmatrix} \frac{4}{3} & 0 & 0 \\ \frac{7}{3} & 1 & 0 \\ 0 & \frac{1}{3} & \frac{2}{3} \end{pmatrix} \\
&= \begin{pmatrix} 37-\frac{4}{3} & 32-0 & 2-0 \\ 21-\frac{7}{3} & 0-1 & 15-0 \\ 5-0 & 0-\frac{1}{3} & 12-\frac{2}{3} \end{pmatrix} = \begin{pmatrix} \frac{107}{3} & 32 & 2 \\ \frac{56}{3} & -1 & 15 \\ 5 & -\frac{1}{3} & \frac{34}{3} \end{pmatrix}.
\end{aligned}
$$

3. $$\left[(3A^T) - \begin{pmatrix} -7 & -2 \\ -6 & 9 \end{pmatrix}^T\right]^T = (3A^T)^T - \left[\begin{pmatrix} -7 & -2 \\ -6 & 9 \end{pmatrix}^T\right]^T$$
$$= 3(A^T)^T - \begin{pmatrix} -7 & -2 \\ -6 & 9 \end{pmatrix}$$
$$= 3A - \begin{pmatrix} -7 & -2 \\ -6 & 9 \end{pmatrix} = \begin{pmatrix} -5 & -10 \\ 33 & 12 \end{pmatrix}.$$

Hence,

$$3A = \begin{pmatrix} -5 & -10 \\ 33 & 12 \end{pmatrix} + \begin{pmatrix} -7 & -2 \\ -6 & 9 \end{pmatrix} = \begin{pmatrix} -5-7 & -10-2 \\ 33-6 & 12+9 \end{pmatrix}$$
$$= \begin{pmatrix} -12 & -22 \\ 33 & 21 \end{pmatrix}.$$

This implies that $A = \frac{1}{3}\begin{pmatrix} -12 & -22 \\ 33 & 21 \end{pmatrix} = \begin{pmatrix} -4 & -\frac{22}{3} \\ 11 & 7 \end{pmatrix}$.

Let

$$C = \left[\frac{1}{2}B + \begin{pmatrix} 6 & 3 \\ 8 & 3 \\ 1 & 4 \end{pmatrix}\right]^T - 3\begin{pmatrix} -5 & 6 \\ 8 & -9 \\ -4 & 2 \end{pmatrix}^T.$$

Then

$$C = \frac{1}{2}B^T + \begin{pmatrix} 6 & 8 & 1 \\ 3 & 3 & 4 \end{pmatrix} - 3\begin{pmatrix} -5 & 8 & -4 \\ 6 & -9 & 2 \end{pmatrix}$$
$$= \frac{1}{2}B^T + \begin{pmatrix} 6+15 & 8-24 & 1+12 \\ 3-18 & 3+27 & 4-6 \end{pmatrix} = \frac{1}{2}B^T + \begin{pmatrix} 21 & -16 & 13 \\ -15 & 30 & -2 \end{pmatrix}$$
$$= \begin{pmatrix} 23 & -16 & 17 \\ -16 & 26.5 & 2 \end{pmatrix}.$$

This implies that

$$\frac{1}{2}B^T = \begin{pmatrix} 23 & -16 & 17 \\ -16 & 26.5 & 2 \end{pmatrix} - \begin{pmatrix} 21 & -16 & 13 \\ -15 & 30 & -2 \end{pmatrix} = \begin{pmatrix} 2 & 0 & 4 \\ -1 & -3.5 & 4 \end{pmatrix}$$

and

$$B^T = 2\begin{pmatrix} 2 & 0 & 4 \\ -1 & -3.5 & 4 \end{pmatrix} = \begin{pmatrix} 4 & 0 & 8 \\ -2 & -7 & 8 \end{pmatrix}.$$

Hence, $B = \begin{pmatrix} 4 & -2 \\ 0 & -7 \\ 8 & 8 \end{pmatrix}$ □

Section 2.7

1. Let $A = \begin{pmatrix} 2 & 1 & 1 \\ 1 & -1 & 2 \end{pmatrix}$, $\overrightarrow{X} = \begin{pmatrix} x_1 \\ x_2 \\ x_3 \end{pmatrix}$, $\overrightarrow{X_1} = \begin{pmatrix} 0 \\ 1 \\ 2 \end{pmatrix}$ and $\overrightarrow{X_2} = \begin{pmatrix} 2 \\ 1 \\ 1 \end{pmatrix}$. Compute $A\overrightarrow{X}$, $A\overrightarrow{X_1}$ and $A\overrightarrow{X_2}$.

2. $A = \begin{pmatrix} -2 & -1 \\ 3 & 1 \\ 0 & 1 \end{pmatrix}$, $\overrightarrow{X} = \begin{pmatrix} -a \\ 2a \end{pmatrix}$ and $\overrightarrow{X_1} = \begin{pmatrix} 1 \\ 2 \end{pmatrix}$.

   Compute $A\overrightarrow{X}$ and $A\overrightarrow{X_1}$.

3. Let $A = \begin{pmatrix} 0 & 1 \\ 1 & -1 \\ 2 & 1 \end{pmatrix}$, $\overrightarrow{X_1} = \begin{pmatrix} 1 \\ 0 \\ -1 \end{pmatrix}$, $\overrightarrow{X_2} = \begin{pmatrix} 1 \\ 0 \end{pmatrix}$ and $\overrightarrow{X_3} = \begin{pmatrix} 1 \\ 1 \\ 1 \\ 2 \end{pmatrix}$.

   Determine whether $A\overrightarrow{X_i}$ is defined for each $i = 1, 2, 3$.

4. Let $A = \begin{pmatrix} 0 & 1 & 0 \\ 1 & 0 & -1 \\ 1 & 1 & 2 \end{pmatrix}$, $\overrightarrow{X} = \begin{pmatrix} 1 \\ 0 \\ -1 \end{pmatrix}$ and $\overrightarrow{Y} = \begin{pmatrix} 3 \\ 0 \\ -1 \end{pmatrix}$.

   Compute $A(\overrightarrow{X} - 3\overrightarrow{Y})$.

5. Let $A = \begin{pmatrix} 1 & 1 & -1 \\ 1 & 0 & 2 \\ 2 & 1 & -1 \end{pmatrix}$ and $\overrightarrow{X} = \begin{pmatrix} 1 \\ -1 \\ 0 \end{pmatrix}$.

   Write $A\overrightarrow{X}$ as a linear combination of the column vectors of A.

Solution.

1. $$A\overrightarrow{X} = \begin{pmatrix} 2 & 1 & 1 \\ 1 & -1 & 2 \end{pmatrix} \begin{pmatrix} x_1 \\ x_2 \\ x_3 \end{pmatrix} = \begin{pmatrix} 2x_1 + x_2 + x_3 \\ x_1 - x_2 + 2x_3 \end{pmatrix}.$$

$$A\overrightarrow{X_1} = \begin{pmatrix} 2 & 1 & 1 \\ 1 & -1 & 2 \end{pmatrix} \begin{pmatrix} 0 \\ 1 \\ 2 \end{pmatrix} = \begin{pmatrix} 2(0) + 1(1) + 1(2) \\ 1(0) + (-1)(1) + 2(2) \end{pmatrix}$$

$$= \begin{pmatrix} 0 + 1 + 2 \\ 0 - 1 + 4 \end{pmatrix} = \begin{pmatrix} 3 \\ 3 \end{pmatrix}.$$

$$A\overrightarrow{X_2} = \begin{pmatrix} 2 & 1 & 1 \\ 1 & -1 & 2 \end{pmatrix} \begin{pmatrix} 2 \\ 1 \\ 1 \end{pmatrix} = \begin{pmatrix} 2(2) + 1(1) + 1(1) \\ 1(2) - 1(1) + 2(1) \end{pmatrix}$$

$$= \begin{pmatrix} 4 + 1 + 1 \\ 2 - 1 + 2 \end{pmatrix} = \begin{pmatrix} 6 \\ 3 \end{pmatrix}.$$

2. $$A\overrightarrow{X} = \begin{pmatrix} -2 & -1 \\ 3 & 1 \\ 0 & 1 \end{pmatrix} \begin{pmatrix} -a \\ 2a \end{pmatrix} = \begin{pmatrix} 2a - 2a \\ -3a + 2a \\ 0 + 2a \end{pmatrix} = \begin{pmatrix} 0 \\ -a \\ 2a \end{pmatrix}.$$

$$A\overrightarrow{X_1} = \begin{pmatrix} -2 & -1 \\ 3 & 1 \\ 0 & 1 \end{pmatrix} \begin{pmatrix} 1 \\ 2 \end{pmatrix} = \begin{pmatrix} -4 \\ 5 \\ 2 \end{pmatrix}.$$

3. $A\overrightarrow{X_1}$ is not defined; $A\overrightarrow{X_2}$ is defined and $A\overrightarrow{X_3}$ is not defined.

4. $$\overrightarrow{X} - 3\overrightarrow{Y} = \begin{pmatrix} 1 \\ 0 \\ -1 \end{pmatrix} - 3 \begin{pmatrix} 3 \\ 0 \\ -1 \end{pmatrix} = \begin{pmatrix} 1 - 9 \\ 0 - 0 \\ -1 + 3 \end{pmatrix} = \begin{pmatrix} -8 \\ 0 \\ 2 \end{pmatrix}.$$

$$A(\overrightarrow{X} - 3\overrightarrow{Y}) = \begin{pmatrix} 0 & 1 & 0 \\ 1 & 0 & -1 \\ 1 & 1 & 2 \end{pmatrix} \begin{pmatrix} -8 \\ 0 \\ 2 \end{pmatrix} = \begin{pmatrix} 0(-8) + 1(0) + 0(2) \\ 1(-8) + 0(0) - 1(2) \\ 1(-8) + 1(0) + 2(2) \end{pmatrix}$$

$$= \begin{pmatrix} 0 \\ -10 \\ -4 \end{pmatrix}.$$

5. $$A\overrightarrow{X} = \begin{pmatrix} 1 & 1 & -1 \\ 1 & 0 & 2 \\ 2 & 1 & -1 \end{pmatrix} \begin{pmatrix} 1 \\ -1 \\ 0 \end{pmatrix} = \begin{pmatrix} 1 \\ 1 \\ 2 \end{pmatrix} - \begin{pmatrix} 1 \\ 0 \\ 1 \end{pmatrix} + 0 \begin{pmatrix} -1 \\ 2 \\ -1 \end{pmatrix}$$ □

Section 2.8

1. Let $A = \begin{pmatrix} 2 & 1 \\ -2 & 3 \end{pmatrix}$ and $B = \begin{pmatrix} 1 & -2 & 1 \\ 3 & 4 & 1 \end{pmatrix}$.

   Find AB and the sizes of A, B and AB.

2. Let $A = \begin{pmatrix} 0 & 1 & -1 \\ 2 & 3 & 1 \end{pmatrix}$ and $B = \begin{pmatrix} 2 & 1 & -1 \\ 1 & 0 & 1 \\ -1 & 2 & 4 \end{pmatrix}$.

   Compute AB and find the sizes of A, B and AB.

3. Let $A = \begin{pmatrix} 1 & 0 & -2 \\ 3 & 2 & -1 \end{pmatrix}$ and $B = \begin{pmatrix} 0 & -1 & 2 & 1 \\ -1 & 2 & 1 & -2 \\ 2 & 0 & 0 & 1 \end{pmatrix}$.

   Compute AB and find the sizes of A, B and AB.

Solution. 1.

$$\begin{aligned} AB &= \begin{pmatrix} 2 & 1 \\ -2 & 3 \end{pmatrix} \begin{pmatrix} 1 & -2 & 1 \\ 3 & 4 & 1 \end{pmatrix} \\ &= \begin{pmatrix} 2(1)+1(3) & 2(-2)+1(4) & 2(1)+1(1) \\ -2(1)+3(3) & -2(-2)+3(4) & -2(1)+3(1) \end{pmatrix} = \begin{pmatrix} 5 & 0 & 3 \\ 7 & 8 & 1 \end{pmatrix}. \end{aligned}$$

The sizes of A, B, AB are $2 \times 2, 2 \times 3, 2 \times 3$, respectively.

2.
$$\begin{aligned} AB &= \begin{pmatrix} 0 & 1 & -1 \\ 2 & 3 & 1 \end{pmatrix} \begin{pmatrix} 2 & 1 & -1 \\ 1 & 0 & 1 \\ -1 & 2 & 4 \end{pmatrix} \\ &= \begin{pmatrix} 0(2)+1(1)-1(-1) & 0(1)+1(0)-1(2) & 0(-1)+1(1)-1(4) \\ 2(2)+3(1)+1(-1) & 2(1)+3(0)+1(2) & 2(-1)+3(1)+1(4) \end{pmatrix} \\ &= \begin{pmatrix} 2 & -2 & -3 \\ 6 & 4 & 5 \end{pmatrix}. \end{aligned}$$

The sizes of A, B, AB are $2 \times 3, 3 \times 3, 2 \times 3$, respectively.

3.
$$AB = \begin{pmatrix} 1 & 0 & -2 \\ 3 & 2 & -1 \end{pmatrix} \begin{pmatrix} 0 & -1 & 2 & 1 \\ -1 & 2 & 1 & -2 \\ 2 & 0 & 0 & 1 \end{pmatrix} = \begin{pmatrix} -4 & -1 & 2 & -1 \\ -4 & 1 & 8 & 2 \end{pmatrix}.$$

The sizes of A, B, AB are $2 \times 3, 3 \times 4, 2 \times 4$, respectively. □

Section 2.9

1. Let $A = \begin{pmatrix} 1 & 1 \\ 1 & 3 \\ 3 & 1 \end{pmatrix}$ and $B = \begin{pmatrix} 2 & 1 \\ -2 & 3 \end{pmatrix}$. Are AB and BA defined?

   If so, compute them. If not, explain why?

2. Let $A = \begin{pmatrix} 2 & -1 \\ -2 & 3 \end{pmatrix}$ and $B = \begin{pmatrix} 0 & -2 \\ 4 & 2 \end{pmatrix}$. Are AB and BA defined?

   If so, compute them. If not, explain why? Is AB equal to BA?

Solution.

1. $$AB = \begin{pmatrix} 1 & 1 \\ 1 & 3 \\ 3 & 1 \end{pmatrix} \begin{pmatrix} 2 & 1 \\ -2 & 3 \end{pmatrix} = \begin{pmatrix} 1(2)+1(-2) & 1(1)+1(3) \\ 1(2)+3(-2) & 1(1)+3(3) \\ 3(2)+1(-2) & 3(1)+1(3) \end{pmatrix} = \begin{pmatrix} 0 & 4 \\ -4 & 10 \\ 4 & 6 \end{pmatrix}.$$

BA is not defined since the size of B is 2×2 and the size of A is 3×2.

2. $$AB = \begin{pmatrix} 2 & -1 \\ -2 & 3 \end{pmatrix} \begin{pmatrix} 0 & -2 \\ 4 & 2 \end{pmatrix} = \begin{pmatrix} 2(0)-1(4) & 2(-2)-1(2) \\ -2(0)+3(4) & -2(-2)+3(2) \end{pmatrix} = \begin{pmatrix} -4 & -6 \\ 12 & 10 \end{pmatrix}.$$

$$BA = \begin{pmatrix} 0 & -2 \\ 4 & 2 \end{pmatrix} \begin{pmatrix} 2 & -1 \\ -2 & 3 \end{pmatrix} = \begin{pmatrix} 0(2)-2(-2) & 0(-1)-2(3) \\ 4(2)+2(-2) & 4(-1)+2(3) \end{pmatrix} = \begin{pmatrix} 4 & -6 \\ 4 & 2 \end{pmatrix}.$$

□

Section 2.10

1. Let

$$A = \begin{pmatrix} 1 & -1 \\ -1 & 2 \end{pmatrix}, B = \begin{pmatrix} 0 & -1 & 2 \\ 1 & 1 & 3 \end{pmatrix} \text{ and } C = \begin{pmatrix} 1 & -1 & 1 \\ 0 & 3 & 2 \\ -2 & 0 & 3 \end{pmatrix}.$$

Show that $A(BC) = (AB)C$.

Solution.

1. $$BC = \begin{pmatrix} 0 & -1 & 2 \\ 1 & 1 & 3 \end{pmatrix} \begin{pmatrix} 1 & -1 & 1 \\ 0 & 3 & 2 \\ -2 & 0 & 3 \end{pmatrix} = \begin{pmatrix} -4 & -3 & 4 \\ -5 & 2 & 12 \end{pmatrix}.$$

$$A(BC) = \begin{pmatrix} 1 & -1 \\ -1 & 2 \end{pmatrix} \begin{pmatrix} -4 & -3 & 4 \\ -5 & 2 & 12 \end{pmatrix} = \begin{pmatrix} 1 & -5 & -8 \\ -6 & 7 & 20 \end{pmatrix}.$$

$$AB = \begin{pmatrix} 1 & -1 \\ -1 & 2 \end{pmatrix} \begin{pmatrix} 0 & -1 & 2 \\ 1 & 1 & 3 \end{pmatrix} = \begin{pmatrix} -1 & -2 & -1 \\ 2 & 3 & 4 \end{pmatrix}.$$

$$(AB)C = \begin{pmatrix} -1 & -2 & -1 \\ 2 & 3 & 4 \end{pmatrix} \begin{pmatrix} 1 & -1 & 1 \\ 0 & 3 & 2 \\ -2 & 0 & 3 \end{pmatrix} = \begin{pmatrix} 1 & -5 & -8 \\ -6 & 7 & 20 \end{pmatrix}.$$

Hence, $A(BC) = (AB)C$. □

Section 2.11

1. Let $A = \begin{pmatrix} 1 & 0 \\ -1 & 1 \end{pmatrix}$. Find A^2, A^3 and A^4.

2. Let $P(x) = 1 - 2x - x^2$ and $A = \begin{pmatrix} -1 & 1 \\ 1 & 2 \end{pmatrix}$. Compute $P(A)$.

3. Let $A = \begin{pmatrix} 2 & 0 & 0 \\ 0 & -2 & 0 \\ 0 & 0 & 3 \end{pmatrix}$. Find A^6.

Solution.

$$1. A^2 = \begin{pmatrix} 1 & 0 \\ -1 & 1 \end{pmatrix}\begin{pmatrix} 1 & 0 \\ -1 & 1 \end{pmatrix} = \begin{pmatrix} 1 & 0 \\ -2 & 1 \end{pmatrix}.$$

$$A^3 = A^2 \cdot A = \begin{pmatrix} 1 & 0 \\ -2 & 1 \end{pmatrix}\begin{pmatrix} 1 & 0 \\ -1 & 1 \end{pmatrix} = \begin{pmatrix} 1 & 0 \\ -3 & 1 \end{pmatrix}.$$

$$A^4 = A^3 \cdot A = \begin{pmatrix} 1 & 0 \\ -3 & 1 \end{pmatrix}\begin{pmatrix} 1 & 0 \\ -1 & 1 \end{pmatrix} = \begin{pmatrix} 1 & 0 \\ -4 & 1 \end{pmatrix}.$$

$$2. \quad P(A) = I - 2A - A^2 = \begin{pmatrix} 1 & 0 \\ 0 & 1 \end{pmatrix} - 2\begin{pmatrix} -1 & 1 \\ 1 & 2 \end{pmatrix} - \begin{pmatrix} -1 & 1 \\ 1 & 2 \end{pmatrix}^2$$

$$= \begin{pmatrix} 1 & 0 \\ 0 & 1 \end{pmatrix} - 2\begin{pmatrix} -1 & 1 \\ 1 & 2 \end{pmatrix} - \begin{pmatrix} -1 & 1 \\ 1 & 2 \end{pmatrix}\begin{pmatrix} -1 & 1 \\ 1 & 2 \end{pmatrix}$$

$$= \begin{pmatrix} 3 & -2 \\ -2 & -3 \end{pmatrix} - \begin{pmatrix} 2 & 1 \\ 1 & 5 \end{pmatrix} = \begin{pmatrix} 1 & -3 \\ -3 & -8 \end{pmatrix}.$$

$$3. \quad A^6 = \begin{pmatrix} 2 & 0 & 0 \\ 0 & -2 & 0 \\ 0 & 0 & 3 \end{pmatrix}^6 = \begin{pmatrix} 2^6 & 0 & 0 \\ 0 & (-2)^6 & 0 \\ 0 & 0 & 3^6 \end{pmatrix} = \begin{pmatrix} 64 & 0 & 0 \\ 0 & 64 & 0 \\ 0 & 0 & 729 \end{pmatrix}.$$

□

Section 2.12

1. Let $A = \begin{pmatrix} 1 & 0 & 1 \\ 2 & 1 & -1 \\ -3 & 2 & -2 \end{pmatrix}$. Use the second row operation to make the numbers 2 and -3 in A to be zero.

2. Let $A = \begin{pmatrix} 1 & 0 & 1 \\ 0 & 1 & -3 \\ 0 & 3 & 1 \end{pmatrix}$. Use the second row operation to make the number 3 in A to be zero.

3. Find a row echelon matrix for each of the following matrices.

$$A = \begin{pmatrix} 1 & 0 & 1 \\ 2 & 1 & -1 \\ -3 & 2 & -2 \end{pmatrix} \quad B = \begin{pmatrix} 0 & 1 & -2 \\ 1 & -1 & 1 \\ -2 & 1 & 0 \end{pmatrix}$$

4. Find the reduced row echelon matrix of each of the following matrices.

$$A = \begin{pmatrix} 2 & -4 & 2 \\ 0 & 1 & -3 \\ 0 & 0 & -6 \end{pmatrix} \quad B = \begin{pmatrix} 1 & 1 & 0 & 1 & 0 \\ 0 & 3 & 6 & -3 & -6 \\ 0 & 0 & 0 & -2 & 2 \\ 0 & 0 & 0 & 0 & 2 \end{pmatrix}$$

$$C = \begin{pmatrix} 1 & 0 & 0 & -1 \\ 1 & -2 & 1 & 1 \\ 2 & -2 & 1 & 2 \end{pmatrix} \quad D = \begin{pmatrix} 0 & 0 & 1 & 1 & -1 \\ -1 & 1 & 2 & -2 & -1 \\ -2 & -2 & 4 & 0 & 0 \\ 2 & -1 & 0 & 1 & 0 \end{pmatrix}$$

$$E = \begin{pmatrix} 2 & 12 & 0 & -1 \\ 3 & 0 & 3 & 2 \\ -3 & 2 & -1 & 6 \end{pmatrix}$$

Solution.

1. $$A = \begin{pmatrix} 1 & 0 & 1 \\ 2 & 1 & -1 \\ -3 & 2 & -2 \end{pmatrix} \xrightarrow[R_1(3)+R_3]{R_1(-2)+R_2} \begin{pmatrix} 1 & 0 & 1 \\ 0 & 1 & -3 \\ 0 & 2 & 1 \end{pmatrix}.$$

2. $$A = \begin{pmatrix} 1 & 0 & 1 \\ 0 & 1 & -3 \\ 0 & 3 & 1 \end{pmatrix} \xrightarrow{R_2(-3)+R_3} \begin{pmatrix} 1 & 0 & 1 \\ 0 & 1 & -3 \\ 0 & 0 & 10 \end{pmatrix}.$$

3. $$A = \begin{pmatrix} 1 & 0 & 1 \\ 2 & 1 & -1 \\ -3 & 2 & -2 \end{pmatrix} \xrightarrow[R_1(3)+R_3]{R_1(-2)+R_2} \begin{pmatrix} 1 & 0 & 1 \\ 0 & 1 & -3 \\ 0 & 2 & 1 \end{pmatrix} \xrightarrow{R_2(-2)+R_3}$$
$$\begin{pmatrix} 1 & 0 & 1 \\ 0 & 1 & -3 \\ 0 & 0 & 7 \end{pmatrix}.$$

$$B = \begin{pmatrix} 0 & 1 & -2 \\ 1 & -1 & 1 \\ -2 & 1 & 0 \end{pmatrix} \xrightarrow{R_{1,2}} \begin{pmatrix} 1 & -1 & 1 \\ 0 & 1 & -2 \\ -2 & 1 & 0 \end{pmatrix} \xrightarrow{R_1(2)+R_2}$$

$$\begin{pmatrix} 1 & -1 & 1 \\ 0 & 1 & -2 \\ 0 & -1 & 2 \end{pmatrix} \xrightarrow{R_2(1)+R_3} \begin{pmatrix} 1 & -1 & 1 \\ 0 & 1 & -2 \\ 0 & 0 & 0 \end{pmatrix}.$$

4. $$A = \begin{pmatrix} 2 & -4 & 2 \\ 0 & 1 & -3 \\ 0 & 0 & -6 \end{pmatrix} \xrightarrow[R_3(-\frac{1}{6})]{R_1(\frac{1}{2})} \begin{pmatrix} 1 & -2 & 1 \\ 0 & 1 & -3 \\ 0 & 0 & 1 \end{pmatrix} \xrightarrow[R_3(3)+R_2]{R_3(-1)+R_1}$$

$$\begin{pmatrix} 1 & -2 & 0 \\ 0 & 1 & 0 \\ 0 & 0 & 1 \end{pmatrix} \xrightarrow{R_2(2)+R_1} \begin{pmatrix} 1 & 0 & 0 \\ 0 & 1 & 0 \\ 0 & 0 & 1 \end{pmatrix}.$$

$$B = \begin{pmatrix} 1 & 1 & 0 & 1 & 0 \\ 0 & 3 & 6 & -3 & -6 \\ 0 & 0 & 0 & -2 & 2 \\ 0 & 0 & 0 & 0 & 2 \end{pmatrix} \xrightarrow[R_4(\frac{1}{2})]{R_2(\frac{1}{3}),\, R_3(-\frac{1}{2})} \begin{pmatrix} 1 & 1 & 0 & 1 & 0 \\ 0 & 1 & 2 & -1 & -2 \\ 0 & 0 & 0 & 1 & -1 \\ 0 & 0 & 0 & 0 & 1 \end{pmatrix}$$

$$\xrightarrow[R_4(1)+R_3]{R_4(2)+R_2} \begin{pmatrix} 1 & 1 & 0 & 1 & 0 \\ 0 & 1 & 2 & -1 & 0 \\ 0 & 0 & 0 & 1 & 0 \\ 0 & 0 & 0 & 0 & 1 \end{pmatrix} \xrightarrow[R_3(1)+R_2]{R_3(-1)+R_1}$$

$$\begin{pmatrix} 1 & 1 & 0 & 0 & 0 \\ 0 & 1 & 2 & 0 & 0 \\ 0 & 0 & 0 & 1 & 0 \\ 0 & 0 & 0 & 0 & 1 \end{pmatrix} \xrightarrow{R_2(-1)+R_1} \begin{pmatrix} 1 & 0 & -2 & 0 & 0 \\ 0 & 1 & 2 & 0 & 0 \\ 0 & 0 & 0 & 1 & 0 \\ 0 & 0 & 0 & 0 & 1 \end{pmatrix}.$$

$$C = \begin{pmatrix} 1 & 0 & 0 & -1 \\ 1 & -2 & 1 & 1 \\ 2 & -2 & 1 & 2 \end{pmatrix} \xrightarrow[R_1(-2)+R_3]{R_1(-1)+R_2} \begin{pmatrix} 1 & 0 & 0 & -1 \\ 0 & -2 & 1 & 2 \\ 0 & -2 & 1 & 4 \end{pmatrix} \xrightarrow{R_2(-1)}$$

$$\begin{pmatrix} 1 & 0 & 0 & -1 \\ 0 & 2 & -1 & -2 \\ 0 & -2 & 1 & 4 \end{pmatrix} \xrightarrow{R_2(1)+R_3} \begin{pmatrix} 1 & 0 & 0 & -1 \\ 0 & 2 & -1 & -2 \\ 0 & 0 & 0 & 2 \end{pmatrix} \xrightarrow{R_3(\frac{1}{2})}$$

$$\begin{pmatrix} 1 & 0 & 0 & -1 \\ 0 & 2 & -1 & -2 \\ 0 & 0 & 0 & 1 \end{pmatrix} \xrightarrow[R_3(2)+R_2]{R_3(1)+R_1} \begin{pmatrix} 1 & 0 & 0 & 0 \\ 0 & 2 & -1 & 0 \\ 0 & 0 & 0 & 1 \end{pmatrix} \xrightarrow{R_2(\frac{1}{2})}$$

$$\begin{pmatrix} 1 & 0 & 0 & 0 \\ 0 & 1 & -\frac{1}{2} & 0 \\ 0 & 0 & 0 & 1 \end{pmatrix}.$$

$$D = \begin{pmatrix} 0 & 0 & 1 & 1 & -1 \\ -1 & 1 & 2 & -2 & -1 \\ -2 & -2 & 4 & 0 & 0 \\ 2 & -1 & 0 & 1 & 0 \end{pmatrix} \xrightarrow{R_{1,2}} \begin{pmatrix} -1 & 1 & 2 & -2 & -1 \\ 0 & 0 & 1 & 1 & -1 \\ -2 & -2 & 4 & 0 & 0 \\ 2 & -1 & 0 & 1 & 0 \end{pmatrix} \xrightarrow{R_1(-1)}$$

$$\begin{pmatrix} 1 & -1 & -2 & 2 & 1 \\ 0 & 0 & 1 & 1 & -1 \\ -2 & -2 & 4 & 0 & 0 \\ 2 & -1 & 0 & 1 & 0 \end{pmatrix} \xrightarrow[R_1(-2)+R_4]{R_1(2)+R_3} \begin{pmatrix} 1 & -1 & -2 & 2 & 1 \\ 0 & 0 & 1 & 1 & -1 \\ 0 & -4 & 0 & 4 & 2 \\ 0 & 1 & 4 & -3 & -2 \end{pmatrix}$$

$$\xrightarrow{R_{2,4}} \begin{pmatrix} 1 & -1 & -2 & 1 & 1 \\ 0 & 1 & 4 & -3 & -2 \\ 0 & -4 & 0 & 2 & 2 \\ 0 & 0 & 1 & 1 & -1 \end{pmatrix} \xrightarrow{R_2(4)+R_3} \begin{pmatrix} 1 & -1 & -2 & 1 & 1 \\ 0 & 1 & 4 & -1 & -2 \\ 0 & 0 & 16 & -10 & -6 \\ 0 & 0 & 1 & 1 & -1 \end{pmatrix}$$

$$\xrightarrow{R_{3,4}} \begin{pmatrix} 1 & -1 & -2 & 1 & 1 \\ 0 & 1 & 4 & -1 & -2 \\ 0 & 0 & 1 & 1 & -1 \\ 0 & 0 & 16 & -10 & -6 \end{pmatrix} \xrightarrow{R_3(-16)+R_4}$$

$$\begin{pmatrix} 1 & -1 & -2 & 1 & 1 \\ 0 & 1 & 4 & -1 & -2 \\ 0 & 0 & 1 & 1 & -1 \\ 0 & 0 & 0 & -26 & 10 \end{pmatrix} \xrightarrow{R_4(-\frac{1}{26})} \begin{pmatrix} 1 & -1 & -2 & 1 & 1 \\ 0 & 1 & 4 & -1 & -2 \\ 0 & 0 & 1 & 1 & -1 \\ 0 & 0 & 0 & 1 & -\frac{5}{13} \end{pmatrix}$$

$$\xrightarrow[\substack{R_4(1)+R_2 \\ R_4(-1)+R_3}]{R_4(-1)+R_1} \begin{pmatrix} 1 & -1 & -2 & 0 & \frac{18}{13} \\ 0 & 1 & 4 & 0 & -\frac{31}{13} \\ 0 & 0 & 1 & 0 & -\frac{8}{13} \\ 0 & 0 & 0 & 1 & -\frac{5}{13} \end{pmatrix} \xrightarrow[R_3(-4)+R_2]{R_3(2)+R_1}$$

$$\begin{pmatrix} 1 & -1 & 0 & 0 & \frac{2}{13} \\ 0 & 1 & 0 & 0 & \frac{1}{13} \\ 0 & 0 & 1 & 0 & -\frac{8}{13} \\ 0 & 0 & 0 & 1 & -\frac{5}{13} \end{pmatrix} \xrightarrow{R_2(1)+R_1} \begin{pmatrix} 1 & 0 & 0 & 0 & \frac{3}{13} \\ 0 & 1 & 0 & 0 & \frac{1}{13} \\ 0 & 0 & 1 & 0 & -\frac{8}{13} \\ 0 & 0 & 0 & 1 & -\frac{5}{13} \end{pmatrix}.$$

$$E = \begin{pmatrix} 2 & 12 & 0 & -1 \\ 3 & 0 & 3 & 2 \\ -3 & 2 & -1 & 6 \end{pmatrix} \xrightarrow{R_2(1)+R_3} \begin{pmatrix} 2 & 12 & 0 & -1 \\ 3 & 0 & 3 & 2 \\ 0 & 2 & 2 & 8 \end{pmatrix}$$

$$\xrightarrow[R_3(\frac{1}{2})]{R_2(-1)+R_1} \begin{pmatrix} -1 & 12 & -3 & -3 \\ 3 & 0 & 3 & 2 \\ 0 & 1 & 1 & 4 \end{pmatrix} \xrightarrow{R_1(3)+R_2}$$

$$\begin{pmatrix} -1 & 12 & -3 & -3 \\ 0 & 36 & -6 & -7 \\ 0 & 1 & 1 & 4 \end{pmatrix} \xrightarrow[R_{2,3}]{R_1(-1)} \begin{pmatrix} 1 & -12 & 3 & 3 \\ 0 & 1 & 1 & 4 \\ 0 & 36 & -6 & -7 \end{pmatrix}$$

$$\xrightarrow{R_2(-36)+R_3} \begin{pmatrix} 1 & -12 & 3 & 3 \\ 0 & 1 & 1 & 4 \\ 0 & 0 & -42 & 151 \end{pmatrix} \xrightarrow[R_3(-\frac{1}{42})]{R_2(12)+R_1} \begin{pmatrix} 1 & 0 & 15 & 51 \\ 0 & 1 & 1 & 4 \\ 0 & 0 & 1 & -\frac{151}{42} \end{pmatrix}$$

$$\xrightarrow[R_3(-1)+R_2]{R_3(-15)+R_1} \begin{pmatrix} 1 & 0 & 0 & \frac{4407}{42} \\ 0 & 1 & 0 & \frac{319}{42} \\ 0 & 0 & 1 & -\frac{151}{42} \end{pmatrix}.$$

□

Section 2.13

1. Find the ranks and nullities of the following matrices.

$$A = \begin{pmatrix} 1 & 0 & 0 & 1 \\ 0 & -1 & 0 & 0 \\ 0 & 0 & 0 & 0 \end{pmatrix} \; B = \begin{pmatrix} 0 & 2 & -1 \\ 2 & -1 & 3 \end{pmatrix} \; C = \begin{pmatrix} 1 & 4 & 2 & 1 \\ 0 & 2 & 1 & 1 \\ 1 & 0 & 1 & 4 \\ 2 & 0 & 1 & -3 \end{pmatrix}$$

Solution. Since A is a row echelon matrix, so $R(A) = 2$ and nullity $(A) = 4 - 2 = 2$.

$$B = \begin{pmatrix} 0 & 2 & -1 \\ 2 & -1 & 3 \end{pmatrix} \xrightarrow{R_{1,2}} \begin{pmatrix} 2 & -1 & 3 \\ 0 & 2 & -1 \end{pmatrix}.$$

$R(B) = 2$ and nullity $(B) = 3 - 2 = 1$

$$C = \begin{pmatrix} 1 & 4 & 2 & 1 \\ 0 & 2 & 1 & 1 \\ 1 & 0 & 1 & 4 \\ 2 & 0 & 1 & -3 \end{pmatrix} \xrightarrow[R_1(-2)+R_4]{R_1(-1)+R_3} \begin{pmatrix} 1 & 4 & 2 & 1 \\ 0 & 2 & 1 & 1 \\ 0 & -4 & -1 & 3 \\ 0 & -8 & -3 & -5 \end{pmatrix} \xrightarrow[R_2(4)+R_4]{R_2(2)+R_3}$$

$$\begin{pmatrix} 1 & 4 & 2 & 1 \\ 0 & 2 & 1 & 1 \\ 0 & 0 & 1 & 5 \\ 0 & 0 & 1 & -1 \end{pmatrix} \xrightarrow{R_3(-1)+R_4} \begin{pmatrix} 1 & 4 & 2 & 1 \\ 0 & 2 & 1 & 1 \\ 0 & 0 & 1 & 5 \\ 0 & 0 & 0 & -6 \end{pmatrix}.$$

$R(C) = 4$ and nullity $(C) = 4 - 4 = 0$ □

Section 2.14

1. Let $A = \begin{pmatrix} \frac{1}{3} & 0 & 0 \\ 0 & \frac{1}{4} & 0 \\ 0 & 0 & 4 \end{pmatrix}$. Find A^{-1}.

2. Let $A = \begin{pmatrix} 3 & -4 \\ 2 & 3 \end{pmatrix}$. Determine if A is invertible. If so, calculate A^{-1}.

3. Let $A = \begin{pmatrix} 2 & 4 \\ -1 & -2 \end{pmatrix}$. Determine if A is invertible. If so, calculate A^{-1}.

4. Let $A = \begin{pmatrix} 1 & x^2 \\ 1 & 1 \end{pmatrix}$. Find all $x \in \mathbb{R}$ such that A is invertible.

5. Let $A = \begin{pmatrix} x & 3 \\ 3 & x \end{pmatrix}$. Find all $x \in \mathbb{R}$ such that A is not invertible.

6. Let $A = \begin{pmatrix} 1 & -3 \\ -4 & 5 \end{pmatrix}$. Use the row operations to determine if A is invertible. If so, calculate A^{-1}.

7. Use the row operations to determine if the following matrix is invertible. If so, calculate its inverse.

$$A = \begin{pmatrix} 1 & -1 & 1 \\ 0 & 1 & -1 \\ 2 & 0 & -1 \end{pmatrix} \qquad B = \begin{pmatrix} 1 & 0 & 1 & 1 \\ 0 & 2 & 1 & 6 \\ -4 & 0 & -3 & -3 \\ 2 & 1 & -3 & 0 \end{pmatrix}$$

8. Let $A = \begin{pmatrix} 1 & 0 \\ 2 & 3 \end{pmatrix}$ and $B = \begin{pmatrix} -1 & 2 \\ 2 & 1 \end{pmatrix}$. Find A^{-1}, B^{-1}, $(AB)^{-1}$ and verify $(AB)^{-1} = B^{-1}A^{-1}$.

9. Let $A = \begin{pmatrix} -1 & 2 \\ 1 & 3 \end{pmatrix}$. Find A^3, A^{-1}, $(A^{-1})^3$ and verify

$$(A^3)^{-1} = (A^{-1})^3.$$

10. Let $A = \begin{pmatrix} -4 & 1 \\ 3 & 1 \end{pmatrix}$. Find A^{-1}, $(A^T)^{-1}$ and verify that

$$(A^T)^{-1} = (A^{-1})^T.$$

11. Let $A = \begin{pmatrix} 2 & 2 \\ 2 & 1 \end{pmatrix}$. Find A^{-1}. Is A^{-1} symmetric.

12. Determine whether the following triangular matrices are invertible.

$$A = \begin{pmatrix} 1 & 0 & 0 \\ 2 & 1 & 0 \\ 3 & 0 & -1 \end{pmatrix} \qquad B = \begin{pmatrix} 0 & 0 & 0 \\ 1 & 1 & 0 \\ 0 & 1 & 2 \end{pmatrix} \qquad C = \begin{pmatrix} 1 & 0 & 0 \\ 0 & 3 & 1 \\ 0 & 0 & 4 \end{pmatrix}$$

$$D = \begin{pmatrix} 5 & 3 & 0 \\ 0 & 1 & 3 \\ 0 & 0 & 0 \end{pmatrix}$$

Solution. 1. Since A is a diagonal matrix, so $A^{-1} = \begin{pmatrix} 3 & 0 & 0 \\ 0 & 4 & 0 \\ 0 & 0 & \frac{1}{4} \end{pmatrix}$.

2. $|A| = \begin{vmatrix} 3 & -4 \\ 2 & 3 \end{vmatrix} = 9 - (-8) = 17 \neq 0$, so A is invertible.

$A^{-1} = \frac{1}{17}\begin{pmatrix} 3 & 4 \\ -2 & 3 \end{pmatrix} = \begin{pmatrix} \frac{3}{17} & \frac{4}{17} \\ -\frac{2}{17} & \frac{3}{17} \end{pmatrix}$.

3. $|A| = \begin{vmatrix} 2 & 4 \\ -1 & -2 \end{vmatrix} = -4 - (-4) = 0$, so A is not invertible.

4. $|A| = \begin{vmatrix} 1 & x^2 \\ 1 & 1 \end{vmatrix} = 1 - x^2$.

Let $|A| = 1 - x^2 = 0$. Then $x^2 = 1$. This implies $x = -1$ or $x = 1$. Hence when $x \neq -1$ and $x \neq 1$, $|A| = 1 - x^2 \neq 0$ and A is invertible.

5. $|A| = \begin{vmatrix} x & 3 \\ 3 & x \end{vmatrix} = x^2 - 9 = 0, x^2 = 9$. This implies $x = -3$ or $x = 3$. Hence, when $x = -3$ or $x = 3$, A is not invertible.

6.

$$(A|I) = \left(\begin{array}{cc|cc} 1 & -3 & 1 & 0 \\ -4 & 5 & 0 & 1 \end{array}\right) \xrightarrow{R_1(4) + R_2} \left(\begin{array}{cc|cc} 1 & -3 & 1 & 0 \\ 0 & -7 & 4 & 1 \end{array}\right) \xrightarrow{R_2(-\frac{1}{7})}$$

$$\left(\begin{array}{cc|cc} 1 & -3 & 1 & 0 \\ 0 & 1 & -\frac{4}{7} & -\frac{1}{7} \end{array}\right) \xrightarrow{R_2(3) + R_1} \left(\begin{array}{cc|cc} 1 & 0 & -\frac{5}{7} & -\frac{3}{7} \\ 0 & 1 & -\frac{4}{7} & -\frac{1}{7} \end{array}\right).$$

Hence, $A^{-1} = \begin{pmatrix} -\frac{5}{7} & -\frac{3}{7} \\ -\frac{4}{7} & -\frac{1}{7} \end{pmatrix}$.

$$7. \quad (A|I) = \left(\begin{array}{ccc|ccc} 1 & -1 & 1 & 1 & 0 & 0 \\ 0 & 1 & -1 & 0 & 1 & 0 \\ 2 & 0 & -1 & 0 & 0 & 1 \end{array}\right) \xrightarrow{R_1(-2) + R_3}$$

$$\left(\begin{array}{ccc|ccc} 1 & -1 & 1 & 1 & 0 & 0 \\ 0 & 1 & -1 & 0 & 1 & 0 \\ 0 & 2 & -3 & -2 & 0 & 1 \end{array}\right) \xrightarrow{R_2(-2) + R_3}$$

$$\left(\begin{array}{ccc|ccc} 1 & -1 & 1 & 1 & 0 & 0 \\ 0 & 1 & -1 & 0 & 1 & 0 \\ 0 & 0 & -1 & -2 & -2 & 1 \end{array}\right) \xrightarrow{R_3(-1)}$$

$$\left(\begin{array}{ccc|ccc} 1 & -1 & 1 & 1 & 0 & 0 \\ 0 & 1 & -1 & 0 & 1 & 0 \\ 0 & 0 & 1 & 2 & 2 & -1 \end{array}\right) \xrightarrow[R_3(1) + R_2]{R_3(-1) + R_1}$$

$$\left(\begin{array}{ccc|ccc} 1 & -1 & 0 & -1 & -2 & 1 \\ 0 & 1 & 0 & 2 & 3 & -1 \\ 0 & 0 & 1 & 2 & 2 & -1 \end{array}\right) \xrightarrow{R_2(1) + R_1}$$

$$\left(\begin{array}{ccc|ccc} 1 & 0 & 0 & 1 & 1 & 0 \\ 0 & 1 & 0 & 2 & 3 & -1 \\ 0 & 0 & 1 & 2 & 2 & -1 \end{array}\right).$$

Hence, $A^{-1}=\begin{pmatrix}1&1&0\\2&3&-1\\2&2&-1\end{pmatrix}$

8. Since $|A|=\begin{vmatrix}1&0\\2&3\end{vmatrix}=3$, $A^{-1}=\frac{1}{3}\begin{pmatrix}3&0\\-2&1\end{pmatrix}=\begin{pmatrix}1&0\\-\frac{2}{3}&\frac{1}{3}\end{pmatrix}$.

Since $|B|=\begin{vmatrix}-1&2\\2&1\end{vmatrix}=-1-4=-5$, $B^{-1}=\frac{1}{-5}\begin{pmatrix}1&-2\\-2&-1\end{pmatrix}=\begin{pmatrix}-\frac{1}{5}&\frac{2}{5}\\\frac{2}{5}&\frac{1}{5}\end{pmatrix}$.

$$AB=\begin{pmatrix}1&0\\2&3\end{pmatrix}\begin{pmatrix}-1&2\\2&1\end{pmatrix}=\begin{pmatrix}-1&2\\4&7\end{pmatrix}.$$

$$(AB|I)=\left(\begin{array}{cc|cc}-1&2&1&0\\4&7&0&1\end{array}\right)\xrightarrow{R_1(4)+R_2}\left(\begin{array}{cc|cc}-1&2&1&0\\0&15&4&1\end{array}\right)\xrightarrow[R_2(\frac{1}{15})]{R_1(-1)}$$
$$\left(\begin{array}{cc|cc}1&-2&-1&0\\0&1&\frac{4}{15}&\frac{1}{15}\end{array}\right)\xrightarrow{R_2(2)+R_1}\left(\begin{array}{cc|cc}1&0&-\frac{7}{15}&\frac{2}{15}\\0&1&\frac{4}{15}&\frac{1}{15}\end{array}\right).$$

So $(AB)^{-1}=\begin{pmatrix}-\frac{7}{15}&\frac{2}{15}\\\frac{4}{15}&\frac{1}{15}\end{pmatrix}$.

$$(2).\quad B^{-1}A^{-1}=\begin{pmatrix}-\frac{1}{5}&\frac{2}{5}\\\frac{2}{5}&\frac{1}{5}\end{pmatrix}\begin{pmatrix}1&0\\-\frac{2}{3}&\frac{1}{3}\end{pmatrix}=\begin{pmatrix}-\frac{1}{5}-\frac{4}{15}&0+\frac{2}{15}\\\frac{2}{5}-\frac{2}{15}&0+\frac{1}{15}\end{pmatrix}$$
$$=\begin{pmatrix}-\frac{7}{15}&\frac{2}{15}\\\frac{4}{15}&\frac{1}{15}\end{pmatrix}.$$

So $(AB)^{-1}=B^{-1}A^{-1}$.

9. $A^2=\begin{pmatrix}-1&2\\1&3\end{pmatrix}\begin{pmatrix}-1&2\\1&3\end{pmatrix}=\begin{pmatrix}3&4\\2&11\end{pmatrix}$

$A^3=A^2\cdot A=\begin{pmatrix}3&4\\2&11\end{pmatrix}\begin{pmatrix}-1&2\\1&3\end{pmatrix}=\begin{pmatrix}1&18\\9&37\end{pmatrix}$

$|A|=\begin{vmatrix}-1&2\\1&3\end{vmatrix}=-3-2=-5$, $A^{-1}=\frac{1}{-5}\begin{pmatrix}3&-2\\-1&-1\end{pmatrix}=\begin{pmatrix}-\frac{3}{5}&\frac{2}{5}\\\frac{1}{5}&\frac{1}{5}\end{pmatrix}$

$$(A^{-1})^2=\begin{pmatrix}-\frac{3}{5}&\frac{2}{5}\\\frac{1}{5}&\frac{1}{5}\end{pmatrix}\begin{pmatrix}-\frac{3}{5}&\frac{2}{5}\\\frac{1}{5}&\frac{1}{5}\end{pmatrix}=\begin{pmatrix}\frac{9}{25}+\frac{2}{25}&-\frac{6}{25}+\frac{2}{25}\\-\frac{3}{25}+\frac{1}{25}&\frac{2}{25}+\frac{1}{25}\end{pmatrix}$$
$$=\begin{pmatrix}\frac{11}{25}&-\frac{4}{25}\\-\frac{2}{5}&\frac{3}{25}\end{pmatrix}.$$

$$(A^{-1})^3=(A^{-1})^2A^{-1}=\begin{pmatrix}\frac{11}{25}&-\frac{4}{25}\\-\frac{2}{5}&\frac{3}{25}\end{pmatrix}\begin{pmatrix}-\frac{3}{5}&\frac{2}{5}\\\frac{1}{5}&\frac{1}{5}\end{pmatrix}$$
$$=\begin{pmatrix}-\frac{33}{125}-\frac{4}{125}&\frac{22}{125}-\frac{4}{125}\\\frac{6}{125}+\frac{3}{125}&-\frac{4}{125}+\frac{3}{125}\end{pmatrix}=\begin{pmatrix}-\frac{37}{125}&\frac{18}{125}\\\frac{9}{125}&-\frac{1}{125}\end{pmatrix}.$$

$|A^3| = \begin{vmatrix} 1 & 18 \\ 9 & 37 \end{vmatrix} = 37 - 162 = -125.$

$(A^3)^{-1} = -\frac{1}{125}\begin{pmatrix} 37 & -18 \\ -9 & 1 \end{pmatrix} = \begin{pmatrix} -\frac{37}{125} & \frac{18}{125} \\ \frac{9}{125} & -\frac{1}{125} \end{pmatrix}.$

so $(A^3)^{-1} = (A^{-1})^3$

10. $|A| = \begin{vmatrix} -4 & 1 \\ 3 & 1 \end{vmatrix} = -7$; $A^{-1} = -\frac{1}{7}\begin{pmatrix} 1 & -1 \\ -3 & -4 \end{pmatrix} = \begin{pmatrix} -\frac{1}{7} & \frac{1}{7} \\ -\frac{3}{7} & \frac{4}{7} \end{pmatrix}$

$A^T = \begin{pmatrix} -4 & 3 \\ 1 & 1 \end{pmatrix}$, $|A^T| = |A| = -7$,

$(A^T)^{-1} = -\frac{1}{7}\begin{pmatrix} 1 & -3 \\ -1 & -4 \end{pmatrix} = \begin{pmatrix} -\frac{1}{7} & \frac{3}{7} \\ \frac{1}{7} & \frac{4}{7} \end{pmatrix}.$

$(A^{-1})^T = \begin{pmatrix} -\frac{1}{7} & -\frac{3}{7} \\ \frac{1}{7} & \frac{4}{7} \end{pmatrix}$, so $(A^T)^{-1} = (A^{-1})^T$

11. $|A| = \begin{vmatrix} 2 & 2 \\ 2 & 1 \end{vmatrix} = 2 - 4 = -2,$

$A^{-1} = -\frac{1}{2}\begin{pmatrix} 1 & -2 \\ -2 & 2 \end{pmatrix} = \begin{pmatrix} -\frac{1}{2} & 1 \\ 1 & -1 \end{pmatrix}$

A^{-1} is symmetric.

12. A,C are invertible since all the entries on the main diagonal are nonzero. B,D are not invertible since there are zero entries on the main diagonals. □

A.3 Determinants

Section 3.1

1. Evaluate each of the following determinants.

$$A = \begin{pmatrix} 1 & 1 \\ 4 & -2 \end{pmatrix} \; B = \begin{pmatrix} -2 & 2 \\ 1 & -3 \end{pmatrix} \; C = \begin{pmatrix} 2 & -1 \\ 3 & 0 \end{pmatrix} \; D = \begin{pmatrix} -2 & 1 \\ -6 & 3 \end{pmatrix}$$

Solution.

1. $|A| = \begin{vmatrix} 1 & 1 \\ 4 & -2 \end{vmatrix} = -2 - 4 = -6$. $|B| = \begin{vmatrix} -2 & 2 \\ 1 & -3 \end{vmatrix} = 6 - 2 = 4.$

$|C| = \begin{vmatrix} 2 & -1 \\ 3 & 0 \end{vmatrix} = 0 - (-3) = 3$. $|D| = \begin{vmatrix} -2 & 1 \\ -6 & 3 \end{vmatrix} = -6 - (-6) = 0.$

□

Section 3.2

1. Calculate each of the following determinants by using (3.1) and (3.2).

$$|A| = \begin{vmatrix} 1 & 2 & 3 \\ -2 & 1 & 2 \\ 3 & -1 & 4 \end{vmatrix} \quad |B| = \begin{vmatrix} 2 & -2 & 4 \\ 4 & 3 & 1 \\ 0 & 1 & 2 \end{vmatrix} \text{ and } |C| = \begin{vmatrix} 0 & -1 & 3 \\ 4 & 1 & 2 \\ 0 & 0 & 1 \end{vmatrix}$$

2. Let $|A| = \begin{vmatrix} 1 & 2 & 3 \\ -2 & 1 & 2 \\ 3 & -1 & 4 \end{vmatrix}$.

   (1) Find $M_{11}, M_{12}, M_{13}, A_{11}, A_{12}, A_{13}$ and compute $|A|$ using cofactors.

   (2) Find $M_{21}, M_{22}, M_{23}, A_{21}, A_{22}, A_{23}$ and compute $|A|$ using cofactors.

   (3) Find $M_{31}, M_{32}, M_{33}, A_{31}, A_{32}, A_{33}$ and compute $|A|$ using cofactors.

Solution. 1. $|A| = \left|\begin{array}{ccc|cc} 1 & 2 & 3 & 1 & 2 \\ -2 & 1 & 2 & -2 & 1 \\ 3 & -1 & 4 & 3 & -1 \end{array}\right| = (4+12+6)-(9-2-16) = 31.$

$|B| = \left|\begin{array}{ccc|cc} 2 & -2 & 4 & 2 & -2 \\ 4 & 3 & 1 & 4 & 3 \\ 0 & 1 & 2 & 0 & 1 \end{array}\right| = (12+0+16)-(0+2-16) = 42.$

$|C| = \left|\begin{array}{ccc|cc} 0 & -1 & 3 & 0 & -1 \\ 4 & 1 & 2 & 4 & 1 \\ 0 & 0 & 1 & 0 & 0 \end{array}\right| = (0+0+0)-(0+0-4) = 4.$

2. (1) $M_{11} = \begin{vmatrix} 1 & 2 \\ -1 & 4 \end{vmatrix} = 4-(-2) = 6.$

$A_{11} = (-1)^{1+1}M_{11} = (-1)^2 M_{11} = M_{11} = 6.$

$M_{12} = \begin{vmatrix} -2 & 2 \\ 3 & 4 \end{vmatrix} = -8-6 = -14.$

$A_{12} = (-1)^{1+2}M_{12} = (-1)^3 M_{12} = -M_{12} = -(-14) = 14.$

$M_{13} = \begin{vmatrix} -2 & 1 \\ 3 & -1 \end{vmatrix} = 2-3 = -1.$

$A_{13} = (-1)^{1+3}M_{13} = (-1)^4 M_{13} = M_{13} = -1.$

$|A| = a_{11}A_{11}+a_{12}A_{12}+a_{13}A_{13} = A_{11}+2A_{12}+3A_{13} = 6+2(14)+3(-1) = 31.$

(2) $M_{21} = \begin{vmatrix} 2 & 3 \\ -1 & 4 \end{vmatrix} = 8-(-3) = 11.$ $A_{11} = (-1)^{2+1}M_{21} = -11.$

$M_{22} = \begin{vmatrix} 1 & 3 \\ 3 & 4 \end{vmatrix} = 4-9 = -5.$ $A_{22} = (-1)^{2+2}M_{22} = M_{22} = -5.$

$M_{23} = \begin{vmatrix} 1 & 2 \\ 3 & -1 \end{vmatrix} = -1-6 = -7.$ $A_{23} = (-1)^{2+3}M_{13} = -M_{23} = 7.$

$|A| = a_{21}A_{21}+a_{22}A_{22}+a_{23}A_{23} = -2(-11)+(-5)+2(7) = 31.$

(3) $M_{31} = \begin{vmatrix} 2 & 3 \\ 1 & 2 \end{vmatrix} = 4 - 3 = 1$. $A_{31} = (-1)^{3+1}M_{31} = M_{31} = 1$.

$M_{32} = \begin{vmatrix} 1 & 3 \\ -2 & 2 \end{vmatrix} = 2 - (-6) = 8$. $A_{32} = (-1)^{3+2}M_{32} = -M_{32} = -8$.

$M_{33} = \begin{vmatrix} 1 & 2 \\ -2 & 1 \end{vmatrix} = -1 - (-4) = 5$. $A_{33} = (-1)^{3+3}M_{33} = M_{33} = 5$.

$|A| = a_{31}A_{31} + a_{32}A_{32} + a_{33}A_{33} = 3(1) - (-8) + 4(5) = 31$. □

Section 3.3

1. Find M_{32}, M_{24}, A_{32}, A_{24}, M_{41}, M_{43}, A_{41}, A_{43} if

$$A = \begin{pmatrix} 1 & -1 & 2 & 3 \\ 2 & 2 & 0 & 3 \\ 1 & 2 & 3 & -1 \\ -3 & 0 & 1 & 6 \end{pmatrix}.$$

2. Find M_{32}, M_{24}, A_{32}, A_{24}, M_{41}, M_{43}, A_{41}, A_{43} if

$$A = \begin{pmatrix} 1 & 0 & 2 & -1 \\ 2 & 0 & 0 & 1 \\ 1 & 0 & 1 & -1 \\ -1 & 0 & 1 & 2 \end{pmatrix}.$$

Solution.

1. $M_{32} = \left|\begin{array}{ccc|cc} 1 & 2 & 3 & 1 & 2 \\ -2 & 0 & 3 & -2 & 0 \\ -3 & 1 & 6 & -3 & 1 \end{array}\right. = (0 - 18 - 6) - (0 + 3 - 24) = -3$.

$M_{24} = \left|\begin{array}{ccc|cc} 1 & -1 & 2 & 1 & -1 \\ 1 & 2 & 3 & 1 & 2 \\ -3 & 0 & 1 & -3 & 0 \end{array}\right. = (2 + 9 + 0) - (-12 + 0 - 1) = 24$.

$A_{32} = (-1)^{3+2}M_{32} = -M_{32} = 3$. $A_{24} = (-1)^{2+4}M_{24} = M_{24} = 24$.

$M_{41} = \left|\begin{array}{ccc|cc} -1 & 2 & 3 & -1 & 2 \\ 2 & 0 & 3 & 2 & 0 \\ 2 & 3 & -1 & 2 & 3 \end{array}\right. = (0 + 12 + 18) - (0 - 9 - 4) = 43$.

$M_{43} = \left|\begin{array}{ccc|cc} 1 & -1 & 3 & 1 & -1 \\ 2 & 2 & 3 & 2 & 2 \\ 1 & 2 & -1 & 1 & 2 \end{array}\right. = (-2 - 3 + 12) - (6 + 6 + 2) = -7$.

$A_{41} = (-1)^{4+1}M_{41} = -M_{41} = -43$. $A_{43} = (-1)^{4+3}M_{43} = -M_{43} = 7$.

2. $M_{32} = \left|\begin{array}{ccc|cc} 1 & 2 & -1 & 1 & 2 \\ 2 & 0 & 1 & 2 & 0 \\ -1 & 1 & 2 & -1 & 1 \end{array}\right. = (0 - 2 - 2) - (0 + 1 + 8) = -13$.

$A_{32} = (-1)^{3+2}M_{32} = -M_{32} = 13$.

$$M_{24} = \left| \begin{array}{ccc|cc} 1 & 0 & 2 & 1 & 0 \\ 1 & 0 & 1 & 1 & 0 \\ -1 & 0 & 1 & -1 & 0 \end{array} \right. = (0+0+0) - (0+0+0) = 0.$$

$A_{24} = (-1)^{2+4} M_{24} = M_{24} = 0.$

$$M_{41} = \left| \begin{array}{ccc|cc} 0 & 2 & -1 & 0 & 2 \\ 0 & 0 & 1 & 0 & 0 \\ 0 & 1 & -1 & 0 & 1 \end{array} \right. = (0+0+0) - (0+0+0) = 0.$$

$A_{41} = (-1)^{4+1} M_{41} = -M_{41} = 0.$

$$M_{43} = \left| \begin{array}{ccc|cc} 1 & 0 & -1 & 1 & 0 \\ 2 & 0 & 1 & 2 & 0 \\ 1 & 0 & -1 & 1 & 0 \end{array} \right. = (0+0+0) - (0+0+0) = 0.$$

$A_{43} = (-1)^{4+3} M_{43} = -M_{43} = 0.$ □

Section 3.4

1. Compute the following determinants by using the expansion of cofactors of each row.

$$|A| = \begin{vmatrix} 1 & 0 & 2 & -1 \\ 2 & 0 & 0 & 1 \\ 1 & 0 & 1 & -1 \\ -1 & 0 & 1 & 2 \end{vmatrix} \qquad |B| = \begin{vmatrix} 1 & -1 & 2 & 3 \\ 2 & 2 & 0 & 3 \\ 1 & 2 & 3 & -1 \\ -3 & 0 & 1 & 6 \end{vmatrix}.$$

2. Let

$$|A| = \begin{vmatrix} 0 & 0 & 0 & 1 \\ 0 & 1 & 0 & 1 \\ 0 & 0 & 1 & 0 \\ 1 & 1 & -1 & 0 \end{vmatrix}.$$

Evaluate $|A|$ by using the expansion of cofactors of a suitable row.

Solution.

1. (*i*) Use row 1 to compute $|A|$.

$|A| = a_{11}A_{11} + a_{12}A_{12} + a_{13}A_{13} + a_{14}A_{14} = A_{11} + 0 + 2A_{13} - A_{14}.$

$$M_{11} = \begin{vmatrix} 0 & 0 & 1 \\ 0 & 1 & -1 \\ 0 & 1 & 2 \end{vmatrix} = 0; \qquad A_{11} = (-1)^{1+1} M_{11} = M_{11} = 0.$$

$$M_{13} = \begin{vmatrix} 2 & 0 & 1 \\ 1 & 0 & -1 \\ -1 & 0 & 2 \end{vmatrix} = 0; \qquad A_{13} = (-1)^{1+3} M_{13} = M_{13} = 0.$$

$$M_{14} = \begin{vmatrix} 2 & 0 & 0 \\ 1 & 0 & 1 \\ -1 & 0 & 1 \end{vmatrix} = 0; \qquad A_{14} = (-1)^{1+4} M_{14} = -M_{14} = 0.$$

Hence, $|A| = 0 + 2(0) - (0) = 0.$

(ii) Use row 2 to compute $|A|$.

$|A| = a_{21}A_{21} + a_{22}A_{22} + a_{23}A_{23} + a_{24}A_{24} = 2A_{21} + 0A_{22} + 0A_{23} + A_{24}$.

$$M_{21} = \begin{vmatrix} 0 & 2 & -1 \\ 0 & 1 & -1 \\ 0 & 1 & 2 \end{vmatrix} = 0; \qquad A_{21} = (-1)^{2+1}M_{21} = 0.$$

$$M_{24} = \begin{vmatrix} 1 & 0 & 2 \\ 1 & 0 & 1 \\ -1 & 0 & 1 \end{vmatrix} = 0; \qquad A_{24} = (-1)^{2+4}M_{24} = 0.$$

Hence, $|A| = 2A_{21} + A_{24} = 2(0) + (0) = 0$.

(iii) Use row 3 to compute $|A|$.

$|A| = a_{31}A_{31} + a_{32}A_{32} + a_{33}A_{33} + a_{34}A_{34} = A_{31} + 0A_{32} + A_{23} - A_{34}$.

$$M_{31} = \begin{vmatrix} 0 & 2 & -1 \\ 0 & 0 & 1 \\ 0 & 1 & 2 \end{vmatrix} = 0; \qquad A_{31} = (-1)^{3+1}M_{31} = 0.$$

$$M_{33} = \begin{vmatrix} 1 & 0 & -1 \\ 2 & 0 & 1 \\ -1 & 0 & 2 \end{vmatrix} = 0; \qquad A_{33} = (-1)^{3+3}M_{33} = 0.$$

$$M_{34} = \begin{vmatrix} 1 & 0 & 2 \\ 2 & 0 & 0 \\ -1 & 0 & 1 \end{vmatrix} = 0; \qquad A_{34} = (-1)^{3+4}M_{34} = 0.$$

$|A| = 0 + 0 - 0 = 0$.

(iv) Use row 4 to compute $|A|$.

$|A| = -A_{41} + 0A_{42} + A_{43} + 2A_{44} = -A_{41} + A_{43} + 2A_{44}$.

$$M_{41} = \begin{vmatrix} 0 & 2 & -1 \\ 0 & 0 & 1 \\ 0 & 1 & -1 \end{vmatrix} = 0; \qquad A_{41} = (-1)^{4+1}M_{41} = 0.$$

$$M_{43} = \begin{vmatrix} 1 & 0 & -1 \\ 2 & 0 & 1 \\ 1 & 0 & -1 \end{vmatrix} = 0; \qquad A_{43} = (-1)^{4+3}M_{43} = 0.$$

$$M_{44} = \begin{vmatrix} 1 & 0 & 2 \\ 2 & 0 & 0 \\ 1 & 0 & 1 \end{vmatrix} = 0; \qquad A_{44} = (-1)^{4+4}M_{44} = 0.$$

Hence, $|A| = -0 + 0 + 2(0) = 0$.

(i) Use row 1 to compute $|B|$.

$|B| = a_{11}A_{11} + a_{12}A_{12} + a_{13}A_{13} + a_{14}A_{14} = A_{11} - A_{12} + 2A_{13} + 3A_{14}$.

$$M_{11} = \begin{vmatrix} 2 & 0 & 3 \\ 2 & 3 & -1 \\ 0 & 1 & 6 \end{vmatrix}\begin{matrix} 2 & 0 \\ 2 & 3 \\ 0 & 1 \end{matrix} = (36 + 0 + 6) - (0 - 2 + 0) = 44;$$

$A_{11} = (-1)^{1+1}M_{11} = M_{11} = 44$.

$$M_{12} = \begin{vmatrix} 2 & 0 & 3 \\ 1 & 3 & -1 \\ -3 & 1 & 6 \end{vmatrix} \begin{matrix} 2 & 0 \\ 1 & 3 \\ -3 & 1 \end{matrix} = (36+0+3)-(-27-2+0) = 68;$$

$A_{12} = (-1)^{1+2} M_{12} = -M_{12} = -68.$

$$M_{13} = \begin{vmatrix} 2 & 2 & 3 \\ 1 & 2 & -1 \\ -3 & 0 & 6 \end{vmatrix} \begin{matrix} 2 & 2 \\ 1 & 2 \\ -3 & 0 \end{matrix} = (24+6+0)-(-18+0+12) = 36;$$

$A_{13} = (-1)^{1+3} M_{13} = M_{13} = 36.$

$$M_{14} = \begin{vmatrix} 2 & 2 & 0 \\ 1 & 2 & 3 \\ -3 & 0 & 1 \end{vmatrix} \begin{matrix} 2 & 2 \\ 1 & 2 \\ -3 & 0 \end{matrix} = (4-18+0)-(0+0+2) = -16;$$

$A_{14} = (-1)^{1+4} M_{14} = -M_{14} = 16.$

Hence, $|B| = 44 - (-68) + 2(36) + 3(16) = 44 + 68 + 72 + 48 = 232.$

(ii) Use row 2 to compute $|B|$.

$|B| = a_{21}A_{21} + a_{22}A_{22} + a_{23}A_{23} + a_{24}A_{24} = 2A_{21} + 2A_{22} + 0A_{23} + 3A_{24} = 2A_{21} + 2A_{22} + 3A_{24}.$

$$M_{21} = \begin{vmatrix} -1 & 2 & 3 \\ 2 & 3 & -1 \\ 0 & 1 & 6 \end{vmatrix} \begin{matrix} -1 & 2 \\ 2 & 3 \\ 0 & 1 \end{matrix} = (-18+0+6)-(0+1+24) = -37;$$

$A_{21} = (-1)^{2+1} M_{21} = -(-37) = 37.$

$$M_{22} = \begin{vmatrix} 1 & 2 & 3 \\ 1 & 3 & -1 \\ -3 & 1 & 6 \end{vmatrix} \begin{matrix} 1 & 2 \\ 1 & 3 \\ -3 & 1 \end{matrix} = (18+6+3)-(-27-1+12) = 43;$$

$A_{22} = (-1)^{2+2} M_{22} = 43.$

$$M_{24} = \begin{vmatrix} 1 & -1 & 2 \\ 1 & 2 & 3 \\ -3 & 0 & 1 \end{vmatrix} \begin{matrix} 1 & -1 \\ 1 & 2 \\ -3 & 0 \end{matrix} = (2+9+0)-(-12+0-1) = 24;$$

$A_{24} = (-1)^{2+4} M_{24} = 24.$

Hence, $|B| = 2(37) + 2(43) + 3(24) = 74 + 86 + 72 = 232.$

(iii) Use row 3 to compute $|B|$.

$|B| = a_{31}A_{31} + a_{32}A_{32} + a_{33}A_{33} + a_{34}A_{34} = A_{31} + 2A_{32} + 3A_{23} - A_{34}.$

$$M_{31} = \begin{vmatrix} -1 & 2 & 3 \\ 2 & 0 & 3 \\ 0 & 1 & 6 \end{vmatrix} \begin{matrix} -1 & 2 \\ 2 & 0 \\ 0 & 1 \end{matrix} = (0+0+6)-(0-3+24) = -15;$$

$A_{31} = (-1)^{3+1} M_{31} = -15.$

$$M_{32} = \begin{vmatrix} -1 & 2 & 3 \\ 2 & 0 & 3 \\ -3 & 1 & 6 \end{vmatrix} \begin{matrix} -1 & 2 \\ 2 & 0 \\ -3 & 1 \end{matrix} = (0-18+6)-(0+3+24) = -39;$$

$A_{32} = (-1)^{3+2} M_{32} = -M_{32} = -(-39) = 39.$

$$M_{33} = \left|\begin{array}{ccc|cc} 1 & -1 & 3 & 1 & -1 \\ 2 & 2 & 3 & 2 & 2 \\ -3 & 0 & 6 & -3 & 0 \end{array}\right| = (12+9+0)-(-18+0-12) = 51;$$

$A_{33} = (-1)^{3+3}M_{33} = 51.$

$$M_{34} = \left|\begin{array}{ccc|cc} 1 & -1 & 2 & 1 & -1 \\ 2 & 2 & 0 & 2 & 2 \\ -3 & 0 & 1 & -3 & 0 \end{array}\right| = (2+0+0)-(-12+0-2) = 16;$$

$A_{34} = (-1)^{3+4}M_{34} = -M_{34} = -16.$

Hence, $|B| = -15 + 2(39) + 3(51) - (-16) = -15 + 78 + 153 + 16 = 232.$

(iv) Use row 4 to compute $|B|$.

$|B| = a_{41}A_{41} + a_{42}A_{42} + a_{43}A_{43} + a_{44}A_{44} = -3A_{41} + 0A_{42} + A_{43} + 6A_{44} = -3A_{41} + A_{43} + 6A_{44}.$

$$M_{41} = \left|\begin{array}{ccc|cc} -1 & 2 & 3 & -1 & 2 \\ 2 & 0 & 3 & 2 & 0 \\ 2 & 3 & -1 & 2 & 3 \end{array}\right| = (0+12+18)-(0-9-4) = 43;$$

$A_{41} = (-1)^{4+1}M_{41} = -M_{41} = -43.$

$$M_{43} = \left|\begin{array}{ccc|cc} 1 & -1 & 3 & 1 & -1 \\ 2 & 2 & 3 & 2 & 2 \\ 1 & 2 & -1 & 1 & 2 \end{array}\right| = (-2-3+12)-(6+6+2) = -7;$$

$A_{43} = (-1)^{4+3}M_{43} = -M_{43} = -(-7) = 7.$

$$M_{44} = \left|\begin{array}{ccc|cc} 1 & -1 & 2 & 1 & -1 \\ 2 & 2 & 0 & 2 & 2 \\ 1 & 2 & 3 & 1 & 2 \end{array}\right| = (6+0+8)-(4+0-6) = 16;$$

$A_{44} = (-1)^{4+4}M_{44} = 0.$

Hence, $|B| = -3(-43) + 7 + 6(16) = 129 + 7 + 96 = 232.$

2. We use the first row to compute $|A|$ since it contains more zeros than other rows.

$|A| = a_{11}A_{11} + a_{12}A_{12} + a_{13}A_{13} + a_{14}A_{14} = 0A_{11} + 0A_{12} + 0A_{13} + A_{14} = A_{14}.$

$$M_{14} = \left|\begin{array}{ccc|cc} 0 & 1 & 0 & 0 & 1 \\ 0 & 0 & 1 & 0 & 0 \\ 1 & 1 & -1 & 1 & 1 \end{array}\right| = (0+1+0)-(0+0+0) = 1;$$

$A_{14} = (-1)^{1+4}M_{14} = -M_{14} = -1.$

Hence, $|A| = -1.$

□

Section 3.5

1. Evaluate each of the following determinants by inspection.

$$|A| = \begin{vmatrix} 1 & 2 & -3 \\ 0 & 1 & 2 \\ -2 & -4 & 6 \end{vmatrix} \quad |B| = \begin{vmatrix} 2 & -2 & 4 \\ 2 & -2 & 4 \\ 0 & 1 & 2 \end{vmatrix} \quad |C| = \begin{vmatrix} 1 & 2 & 3 \\ -2 & 1 & 2 \\ 1 & -\frac{1}{2} & -1 \end{vmatrix}$$

$$|D| = \begin{vmatrix} 1 & -1 & 3 & 5 & 0 \\ 0 & 0 & 0 & 0 & 0 \\ 0 & 0 & 1 & 2 & 7 \\ 5 & 6 & 3 & -4 & 6 \\ 2 & -2 & 6 & 10 & 0 \end{vmatrix} \quad |E| = \begin{vmatrix} 1 & -1 & 3 & 5 & 0 \\ 0 & 3 & 0 & 0 & 0 \\ 0 & 0 & -1 & 2 & 7 \\ 0 & 0 & 0 & -4 & 6 \\ 0 & 0 & 0 & 0 & 2 \end{vmatrix}$$

$$|F| = \begin{vmatrix} 1 & 0 & 0 & 0 \\ -2 & 2 & 0 & 0 \\ -1 & 6 & -6 & 0 \\ 4 & 0 & 2 & -1 \end{vmatrix} \quad |G| = \begin{vmatrix} -2 & 0 & 0 & 0 \\ 0 & 3 & 0 & 0 \\ 0 & 0 & -1 & 0 \\ 0 & 0 & 0 & 2 \end{vmatrix}$$

Solution. 2. $|A| = 0$ since row 1 is proportional to row 3.

$|B| = 0$ since $row1$ equals row 2.

$|C| = 0$ since row 2 is proportional to row 3

$|D| = 0$ since row 2 is a zero row.

$|E| = 1(3)(-1)(-4)2 = 24.$

$|F| = 1(2)(-6)(-1) = 12.$

$|G| = (-2)(3)(-1)(2) = 12.$ □

Section 3.6

1. Use row operations to evaluate each of the following determinants.

$$|A| = \begin{vmatrix} 1 & 2 & -3 & -2 \\ 2 & 1 & 2 & 0 \\ -2 & -3 & 1 & -2 \\ 1 & -1 & 2 & 0 \end{vmatrix} \quad |B| = \begin{vmatrix} 2 & 1 & 4 \\ 1 & -2 & 4 \\ 0 & 2 & 4 \end{vmatrix}$$

$$|C| = \begin{vmatrix} 0 & -1 & 3 & 5 & 0 \\ 1 & -1 & 2 & 4 & -1 \\ 0 & 0 & 1 & 2 & 7 \\ 1 & 1 & -2 & 3 & 1 \\ 2 & -2 & 6 & 10 & 0 \end{vmatrix} \quad |D| = \begin{vmatrix} 0 & 2 & 3 \\ -2 & 1 & 2 \\ 3 & -1 & -1 \end{vmatrix}$$

$$|E| = \begin{vmatrix} 1 & -1 & 3 & 5 & 0 \\ 1 & 2 & 0 & 0 & 0 \\ -2 & 0 & -1 & 2 & 1 \\ 3 & 0 & 0 & -6 & 6 \\ 2 & 0 & 0 & 0 & 1 \end{vmatrix} \quad |F| = \begin{vmatrix} 0 & 0 & 0 & 0 \\ -2 & 2 & 0 & 0 \\ 0 & 6 & -6 & 0 \\ 4 & 0 & 2 & -1 \end{vmatrix}$$

$$|G| = \begin{vmatrix} -2 & 0 & 0 & 0 \\ 3 & 3 & 0 & 1 \\ 0 & 1 & -1 & 0 \\ -1 & 0 & 0 & 2 \end{vmatrix}$$

Solution.

$$|A| = \begin{vmatrix} 1 & 2 & -3 & -2 \\ 2 & 1 & 2 & 0 \\ -2 & -3 & 1 & -2 \\ 1 & -1 & 2 & 0 \end{vmatrix} \underset{\substack{R_1(2)+R_3 \\ R_1(-1)+R_4}}{\overset{R_1(-2)+R_2}{=\!=\!=\!=}} \begin{vmatrix} 1 & 2 & -3 & -2 \\ 0 & -3 & 8 & 4 \\ 0 & -1 & -5 & -6 \\ 0 & -3 & 5 & 2 \end{vmatrix}$$

$$\overset{R_{2,3}}{=\!=} - \begin{vmatrix} 1 & 2 & -3 & -2 \\ 0 & -1 & -5 & -6 \\ 0 & -3 & 8 & 4 \\ 0 & -3 & 5 & 2 \end{vmatrix} \underset{R_2(-3)+R_4}{\overset{R_2(-3)+R_3}{=\!=\!=\!=}} - \begin{vmatrix} 1 & 2 & -3 & -2 \\ 0 & -1 & -5 & -6 \\ 0 & 0 & 23 & 22 \\ 0 & 0 & 20 & 20 \end{vmatrix}$$

$$\overset{R_{3,4}}{=\!=\!=} \begin{vmatrix} 1 & 2 & -3 & -2 \\ 0 & -1 & -5 & -6 \\ 0 & 0 & 20 & 20 \\ 0 & 0 & 23 & 22 \end{vmatrix} \overset{R_3(-\frac{23}{20})+R_4}{=\!=\!=\!=} \begin{vmatrix} 1 & 2 & -3 & -2 \\ 0 & -1 & -5 & -6 \\ 0 & 0 & 20 & 20 \\ 0 & 0 & 0 & -1 \end{vmatrix} = 20.$$

$$|B| = \begin{vmatrix} 2 & 1 & 4 \\ 1 & -2 & 4 \\ 0 & 2 & 4 \end{vmatrix} \overset{R_{1,2}}{=\!=\!=} - \begin{vmatrix} 1 & -2 & 4 \\ 2 & 1 & 4 \\ 0 & 2 & 4 \end{vmatrix} \overset{R_1(-2)+R_2}{=\!=\!=\!=} - \begin{vmatrix} 1 & -2 & 4 \\ 0 & 5 & -4 \\ 0 & 2 & 4 \end{vmatrix}$$

$$\overset{R_3(\frac{1}{2})}{=\!=} (-2) \begin{vmatrix} 1 & -2 & 4 \\ 0 & 5 & -4 \\ 0 & 1 & 2 \end{vmatrix} \overset{R_{2,3}}{=\!=\!=} 2 \begin{vmatrix} 1 & -2 & 4 \\ 0 & 1 & 2 \\ 0 & 5 & -4 \end{vmatrix} \overset{R_2(-5)+R_3}{=\!=\!=\!=}$$

$$2 \begin{vmatrix} 1 & -2 & 4 \\ 0 & 1 & 2 \\ 0 & 0 & -14 \end{vmatrix} = 2(-14) = -28.$$

$$|C| = \begin{vmatrix} 0 & -1 & 3 & 5 & 0 \\ 1 & -1 & 2 & 4 & -1 \\ 0 & 0 & 1 & 2 & 7 \\ 1 & 1 & -2 & 3 & 1 \\ 2 & -2 & 6 & 10 & 0 \end{vmatrix} \overset{R_{1,2}}{=\!=\!=} - \begin{vmatrix} 1 & -1 & 2 & 4 & -1 \\ 0 & -1 & 3 & 5 & 0 \\ 0 & 0 & 1 & 2 & 7 \\ 1 & 1 & -2 & 3 & 1 \\ 2 & -2 & 6 & 10 & 0 \end{vmatrix} \overset{R_5(\frac{1}{2})}{=\!=}$$

$$-2 \begin{vmatrix} 1 & -1 & 2 & 4 & -1 \\ 0 & -1 & 3 & 5 & 0 \\ 0 & 0 & 1 & 2 & 7 \\ 1 & 1 & -2 & 3 & 1 \\ 1 & -1 & 3 & 5 & 0 \end{vmatrix} \underset{R_1(-1)+R_5}{\overset{R_1(-1)+R_4}{=\!=\!=\!=}} -2 \begin{vmatrix} 1 & -1 & 2 & 4 & -1 \\ 0 & -1 & 3 & 5 & 0 \\ 0 & 0 & 1 & 2 & 7 \\ 0 & 2 & -1 & -1 & 2 \\ 0 & 0 & 1 & 1 & 1 \end{vmatrix}$$

$$\overset{R_2(2)+R_4}{=\!=\!=\!=} -2 \begin{vmatrix} 1 & -1 & 2 & 4 & -1 \\ 0 & -1 & 3 & 5 & 0 \\ 0 & 0 & 1 & 2 & 7 \\ 0 & 0 & 5 & 9 & 2 \\ 0 & 0 & 1 & 1 & 1 \end{vmatrix} \underset{R_3(-1)+R_5}{\overset{R_3(-5)+R_4}{=\!=\!=\!=}}$$

$$-2 \begin{vmatrix} 1 & -1 & 2 & 4 & -1 \\ 0 & -1 & 3 & 5 & 0 \\ 0 & 0 & 1 & 2 & 7 \\ 0 & 0 & 0 & -1 & -33 \\ 0 & 0 & 0 & -1 & -6 \end{vmatrix} \overset{R_4(-1)+R_3}{=\!=\!=\!=} (-2) \begin{vmatrix} 1 & -1 & 2 & 4 & -1 \\ 0 & -1 & 3 & 5 & 0 \\ 0 & 0 & 1 & 2 & 7 \\ 0 & 0 & 0 & -1 & -33 \\ 0 & 0 & 0 & 0 & -27 \end{vmatrix}$$

$$= -2(-27) = 54.$$

$$|D| = \begin{vmatrix} 0 & 2 & 3 \\ -2 & 1 & 2 \\ 3 & -1 & -1 \end{vmatrix} \overset{R_2(1)+R_3}{=\!=} \begin{vmatrix} 0 & 2 & 3 \\ -2 & 1 & 2 \\ 1 & 0 & 1 \end{vmatrix} \overset{R_{1,3}}{=\!=} -\begin{vmatrix} 1 & 0 & 1 \\ -2 & 1 & 2 \\ 0 & 2 & 3 \end{vmatrix}$$

$$\overset{R_1(2)+R_3}{=\!=} -\begin{vmatrix} 1 & 0 & 1 \\ 0 & 1 & 4 \\ 0 & 0 & -5 \end{vmatrix} = -(-5) = 5.$$

$$|E| = \begin{vmatrix} 1 & -1 & 3 & 5 & 0 \\ 1 & 2 & 0 & 0 & 0 \\ -2 & 0 & -1 & 2 & 1 \\ 3 & 0 & 0 & -6 & 6 \\ 2 & 0 & 0 & 0 & 1 \end{vmatrix} \underset{\substack{R_1(2)+R_3 \\ R_1(-3)+R_4 \\ R_1(-2)+R_5}}{\overset{R_1(-1)+R_2}{=\!=}} \begin{vmatrix} 1 & -1 & 3 & 5 & 0 \\ 0 & 3 & -3 & -5 & 0 \\ 0 & -2 & 5 & 12 & 1 \\ 0 & 3 & -9 & -21 & 6 \\ 0 & 2 & -6 & -10 & 1 \end{vmatrix}$$

$$\underset{R_3(1)+R_5}{\overset{R_2(-1)+R_4}{=\!=}} \begin{vmatrix} 1 & -1 & 3 & 5 & 0 \\ 0 & 3 & -3 & -5 & 0 \\ 0 & -2 & 5 & 12 & 1 \\ 0 & 0 & -6 & -16 & 6 \\ 0 & 0 & -1 & 2 & 2 \end{vmatrix} \underset{R_4(-\frac{1}{2})}{\overset{R_3(1)+R_2}{=\!=}}$$

$$-2\begin{vmatrix} 1 & -1 & 3 & 5 & 0 \\ 0 & 1 & 2 & 7 & 1 \\ 0 & -2 & 5 & 12 & 1 \\ 0 & 0 & 3 & 8 & -3 \\ 0 & 0 & -1 & 2 & 2 \end{vmatrix} \underset{R_5(3)+R_4}{\overset{R_1(2)+R_3}{=\!=}} -2\begin{vmatrix} 1 & -1 & 3 & 5 & 0 \\ 0 & 1 & 2 & 7 & 1 \\ 0 & 0 & 9 & 26 & 3 \\ 0 & 0 & 0 & 14 & 3 \\ 0 & 0 & -1 & 2 & 2 \end{vmatrix}$$

$$\overset{R_5(9)+R_3}{=\!=} -2\begin{vmatrix} 1 & -1 & 3 & 5 & 0 \\ 0 & 1 & 2 & 7 & 1 \\ 0 & 0 & 0 & 44 & 21 \\ 0 & 0 & 0 & -1 & 3 \\ 0 & 0 & -1 & 2 & 2 \end{vmatrix} \overset{R_{3,5}}{=\!=}$$

$$2\begin{vmatrix} 1 & -1 & 3 & 5 & 0 \\ 0 & 1 & 2 & 7 & 1 \\ 0 & 0 & -1 & 2 & 2 \\ 0 & 0 & 0 & -1 & 3 \\ 0 & 0 & 0 & 44 & 21 \end{vmatrix} \overset{R_4(-44)+R_5}{=\!=} 2\begin{vmatrix} 1 & -1 & 3 & 5 & 0 \\ 0 & 1 & 2 & 7 & 1 \\ 0 & 0 & -1 & 2 & 2 \\ 0 & 0 & 0 & -1 & 3 \\ 0 & 0 & 0 & 0 & 153 \end{vmatrix}$$

$$= 2(153) = 306.$$

$$|F| = \begin{vmatrix} 0 & 0 & 0 & 0 \\ -2 & 2 & 0 & 0 \\ 0 & 6 & -6 & 0 \\ 4 & 0 & 2 & -1 \end{vmatrix} \overset{R_2(2)+R_4}{=\!=} \begin{vmatrix} 0 & 0 & 0 & 0 \\ -2 & 2 & 0 & 0 \\ 0 & 6 & -6 & 0 \\ 0 & 4 & 2 & -1 \end{vmatrix} \underset{R_3(\frac{1}{6})}{\overset{R_2(-\frac{1}{2})}{=\!=}}$$

$$(-2)(6)\begin{vmatrix} 0 & 0 & 0 & 0 \\ 1 & 1 & 0 & 0 \\ 0 & 1 & -1 & 0 \\ 0 & 4 & 2 & -1 \end{vmatrix} \underset{R_3(-4)+R_4}{\overset{R_{1,2}}{=\!=}} 12\begin{vmatrix} 1 & 1 & 0 & 0 \\ 0 & 0 & 0 & 0 \\ 0 & 1 & -1 & 0 \\ 0 & 0 & 6 & -1 \end{vmatrix} \overset{R_{2,3}}{=\!=}$$

$$-12\begin{vmatrix} 1 & 1 & 0 & 0 \\ 0 & 1 & -1 & 0 \\ 0 & 0 & 0 & 0 \\ 0 & 0 & 6 & -1 \end{vmatrix} \overset{R_{3,4}}{=\!=} 12\begin{vmatrix} 1 & 1 & 0 & 0 \\ 0 & 1 & -1 & 0 \\ 0 & 0 & 6 & -1 \\ 0 & 0 & 0 & 0 \end{vmatrix} = 12(0) = 0.$$

$$|G| = \begin{vmatrix} -2 & 0 & 0 & 0 \\ 3 & 3 & 0 & 1 \\ 0 & 1 & -1 & 0 \\ -1 & 0 & 0 & 2 \end{vmatrix} \overset{R_1(-\frac{1}{2})}{=\!=\!=} (-2) \begin{vmatrix} 1 & 0 & 0 & 0 \\ 3 & 3 & 0 & 1 \\ 0 & 1 & -1 & 0 \\ -1 & 0 & 0 & 2 \end{vmatrix} \overset{R_1(-3)+R_2}{\underset{R_1(1)+R_4}{=\!=\!=}}$$

$$-2 \begin{vmatrix} 1 & 0 & 0 & 0 \\ 0 & 3 & 0 & 1 \\ 0 & 1 & -1 & 0 \\ 0 & 0 & 0 & 2 \end{vmatrix} \overset{R_{2,3}}{=\!=\!=} 2 \begin{vmatrix} 1 & 0 & 0 & 0 \\ 0 & 1 & -1 & 0 \\ 0 & 3 & 0 & 1 \\ 0 & 0 & 0 & 2 \end{vmatrix} \overset{R_2(-3)+R_3}{=\!=\!=}$$

$$2 \begin{vmatrix} 1 & 0 & 0 & 0 \\ 0 & 1 & -1 & 0 \\ 0 & 0 & 3 & 1 \\ 0 & 0 & 0 & 2 \end{vmatrix} = 12.$$

□

Section 3.7

1. Evaluate each of the following determinants by using its transpose.

$$|A| = \begin{vmatrix} 1 & 0 & 0 & 1 \\ 2 & 7 & 0 & 2 \\ 0 & 6 & 3 & 0 \\ 5 & 3 & 1 & 2 \end{vmatrix} \qquad |B| = \begin{vmatrix} 1 & 0 & 0 & 0 & 0 \\ 2 & 7 & 0 & 2 & 1 \\ 0 & 6 & 3 & 0 & 0 \\ 1 & 3 & 1 & 2 & -1 \\ 0 & 6 & 1 & 2 & 1 \end{vmatrix}$$

Solution.

$$|A| = |A^T| = \begin{vmatrix} 1 & 2 & 0 & 5 \\ 0 & 7 & 6 & 3 \\ 0 & 0 & 3 & 1 \\ 1 & 2 & 0 & 2 \end{vmatrix} \overset{R_1(-1)+R_4}{=\!=\!=} \begin{vmatrix} 1 & 2 & 0 & 5 \\ 0 & 7 & 6 & 3 \\ 0 & 0 & 3 & 1 \\ 0 & 0 & 0 & -3 \end{vmatrix} = -63.$$

$$|B| = |B^T| = \begin{vmatrix} 1 & 2 & 0 & 1 & 0 \\ 0 & 7 & 6 & 3 & 6 \\ 0 & 0 & 3 & 1 & 1 \\ 0 & 2 & 0 & 2 & 2 \\ 0 & 1 & 0 & -1 & 1 \end{vmatrix} \overset{R_{2,5}}{=\!=\!=} - \begin{vmatrix} 1 & 2 & 0 & 1 & 0 \\ 0 & 1 & 0 & -1 & 1 \\ 0 & 0 & 3 & 1 & 1 \\ 0 & 2 & 0 & 2 & 2 \\ 0 & 7 & 6 & 3 & 6 \end{vmatrix} \overset{R_2(-2)+R_4}{\underset{R_2(-7)+R_5}{=\!=\!=}}$$

$$
-\begin{vmatrix} 1 & 2 & 0 & 1 & 0 \\ 0 & 1 & 0 & -1 & 1 \\ 0 & 0 & 3 & 1 & 1 \\ 0 & 0 & 0 & 4 & 0 \\ 0 & 0 & 6 & 10 & -1 \end{vmatrix} \xlongequal{R_3(-2)+R_5} - \begin{vmatrix} 1 & 2 & 0 & 1 & 0 \\ 0 & 1 & 0 & -1 & 1 \\ 0 & 0 & 3 & 1 & 1 \\ 0 & 0 & 0 & 4 & 0 \\ 0 & 0 & 0 & 8 & -3 \end{vmatrix}
$$

$$
\xlongequal{R_4(-2)+R_5} - \begin{vmatrix} 1 & 2 & 0 & 1 & 0 \\ 0 & 1 & 0 & -1 & 1 \\ 0 & 0 & 3 & 1 & 1 \\ 0 & 0 & 0 & 4 & 0 \\ 0 & 0 & 0 & 0 & -3 \end{vmatrix} = 36.
$$

□

Section 3.8

1. Let A be a 4×4 matrix. Assume that $|A| = 2$. Compute each of the following determinants.

   (1) $|-3A|$ (2) $|A^{-1}|$ (3) $|A^T|$ (4) $|A^3|$ (5) $|(2A^{-1})^T|$ (6) $|(2(-A)^T)^{-1}|$

2. Determine whether the given matrix is invertible.

$$
A = \begin{pmatrix} 1 & 0 & 0 & 0 \\ 2 & 1 & 2 & 0 \\ -2 & -3 & 1 & -2 \\ 1 & -1 & 0 & 0 \end{pmatrix} \quad B = \begin{pmatrix} 2 & 1 & 4 \\ 1 & -2 & 4 \\ -2 & -1 & -4 \end{pmatrix}
$$

$$
C = \begin{pmatrix} 0 & -1 & 3 & 5 & 2 \\ 1 & -1 & 2 & 4 & -1 \\ 0 & 0 & 1 & 2 & 7 \\ 1 & 1 & -2 & 3 & 1 \\ 0 & -2 & 6 & 10 & 4 \end{pmatrix} \quad D = \begin{pmatrix} 0 & 1 & 3 \\ -2 & 1 & 2 \\ 1 & -1 & -1 \end{pmatrix}
$$

$$
E = \begin{pmatrix} 1 & -1 & 3 & 5 & 0 \\ 0 & 0 & 0 & 0 & 0 \\ -2 & 0 & -1 & 2 & 1 \\ 3 & 0 & 0 & -6 & 6 \\ 2 & 0 & 0 & 0 & 1 \end{pmatrix} \quad F = \begin{pmatrix} 1 & 0 & 0 & 0 \\ -2 & 2 & 0 & 0 \\ 0 & 6 & -6 & 0 \\ 4 & 0 & 2 & -1 \end{pmatrix}
$$

3. Find all real numbers x such that the following matrix is invertible.

$$
A = \begin{pmatrix} x & 0 & 0 & 0 \\ 2 & x & 2 & 0 \\ -2 & -3 & 1 & -2 \\ 1 & 0 & 0 & 1 \end{pmatrix} \quad B = \begin{pmatrix} 2 & x & 4 \\ 1 & -2 & 4 \\ -2 & -1 & x \end{pmatrix}
$$

4. Let $A = \begin{pmatrix} 2 & 1 & 3 \\ 1 & 0 & 3 \\ 1 & 4 & 7 \end{pmatrix}$ and $B = \begin{pmatrix} 2 & 1 & 3 \\ 1 & 0 & 3 \\ 0 & 1 & -1 \end{pmatrix}$. Compute $|A+B|$.

5. Compute $|A+B|$ if

$$A = \begin{pmatrix} 1 & 0 & 0 & 2 \\ -1 & 0 & 1 & 2 \\ 0 & 1 & 4 & 3 \\ -2 & 2 & 0 & 1 \end{pmatrix} \text{ and } B = \begin{pmatrix} 1 & 0 & 0 & 2 \\ 3 & 0 & 0 & 1 \\ 0 & 1 & 4 & 3 \\ -2 & 2 & 0 & 1 \end{pmatrix}.$$

Solution.

1. $(1)|-3A| = (-3)^4|A| = 81(2) = 162.$

   $(2)|A^{-1}| = \frac{1}{|A|} = \frac{1}{2}.$

   $(3)|A^T| = |A| = 2.$

   $(4)|A^3| = |A|^3 = 2^3 = 8.$

   $(5)|(2A^{-1})^T| = |2A^{-1}| = 2^4|A^{-1}| = \frac{16}{|A|} = \frac{16}{2} = 8.$

   $(6)|(2(-A)^T)^{-1}| = \frac{1}{|2(-A)^T|} = \frac{1}{2^4|(-A)^T|} = \frac{1}{16|-A|} = \frac{1}{16(-1)^4|A|}$

   $= \frac{1}{32}.$

2. $$|A| = \begin{vmatrix} 1 & 0 & 0 & 0 \\ 2 & 1 & 2 & 0 \\ -2 & -3 & 1 & -2 \\ 1 & -1 & 0 & 0 \end{vmatrix} = \begin{vmatrix} 1 & 2 & -2 & 1 \\ 0 & 1 & -3 & -1 \\ 0 & 2 & 1 & 0 \\ 0 & 0 & -2 & 0 \end{vmatrix} \overset{R_2(-2)+R_3}{=\!=\!=}$$

   $$\begin{vmatrix} 1 & 2 & -2 & 1 \\ 0 & 1 & -3 & -1 \\ 0 & 0 & 7 & 2 \\ 0 & 0 & -2 & 0 \end{vmatrix} \overset{R_{3,4}}{=\!=\!=} - \begin{vmatrix} 1 & 2 & -2 & 1 \\ 0 & 1 & -3 & -1 \\ 0 & 0 & -2 & 0 \\ 0 & 0 & 7 & 2 \end{vmatrix} \overset{R_3(\frac{7}{2})+R_4}{=\!=\!=}$$

   $$- \begin{vmatrix} 1 & 2 & -2 & 1 \\ 0 & 1 & -3 & -1 \\ 0 & 0 & -2 & 0 \\ 0 & 0 & 0 & 2 \end{vmatrix} = -(-4) = 4 \neq 0,$$

so A is invertible.

   Since $|B| = \begin{vmatrix} 2 & 1 & 4 \\ 1 & -2 & 4 \\ -2 & -1 & -4 \end{vmatrix} \overset{R_1(1)+R_3}{=\!=\!=} \begin{vmatrix} 2 & 1 & 4 \\ 1 & -2 & 4 \\ 0 & 0 & 0 \end{vmatrix} = 0,$

   so B is not invertible.

$$|C| = \begin{vmatrix} 0 & -1 & 3 & 5 & 2 \\ 1 & -1 & 2 & 4 & -1 \\ 0 & 0 & 1 & 2 & 7 \\ 1 & 1 & -2 & 3 & 1 \\ 0 & -2 & 6 & 10 & 4 \end{vmatrix} \underline{\underline{R_2(-1)+R_4}} \begin{vmatrix} 0 & -1 & 3 & 5 & 2 \\ 1 & -1 & 2 & 4 & -1 \\ 0 & 0 & 1 & 2 & 7 \\ 0 & 2 & -4 & -1 & 2 \\ 0 & -2 & 6 & 10 & 4 \end{vmatrix}$$

$$\underset{R_4(1)+R_5}{\overset{R_{1,2}}{=\!=}} - \begin{vmatrix} 1 & -1 & 2 & 4 & -1 \\ 0 & -1 & 3 & 5 & 2 \\ 0 & 0 & 1 & 2 & 7 \\ 0 & 2 & -4 & -1 & 2 \\ 0 & 0 & 2 & 9 & 6 \end{vmatrix} \underline{\underline{R_2(2)+R_4}}$$

$$- \begin{vmatrix} 1 & -1 & 2 & 4 & -1 \\ 0 & -1 & 3 & 5 & 2 \\ 0 & 0 & 1 & 2 & 7 \\ 0 & 0 & 2 & 9 & 6 \\ 0 & 0 & 2 & 9 & 6 \end{vmatrix} \underset{R_3(-2)+R_5}{\overset{R_3(-2)+R_4}{=\!=}} - \begin{vmatrix} 1 & -1 & 2 & 4 & -1 \\ 0 & -1 & 3 & 5 & 2 \\ 0 & 0 & 1 & 2 & 7 \\ 0 & 0 & 0 & 5 & -8 \\ 0 & 0 & 0 & 5 & -8 \end{vmatrix}$$

$$\underline{\underline{R_4(-1)+R_5}} - \begin{vmatrix} 1 & -1 & 2 & 4 & -1 \\ 0 & -1 & 3 & 5 & 2 \\ 0 & 0 & 1 & 2 & 7 \\ 0 & 0 & 0 & 5 & -8 \\ 0 & 0 & 0 & 0 & 0 \end{vmatrix} = 0,$$

so C is not invertible.

$$|D| = \begin{vmatrix} 0 & 1 & 3 \\ -2 & 1 & 2 \\ 1 & -1 & -1 \end{vmatrix} \underline{\underline{R_{1,3}}} - \begin{vmatrix} 1 & -1 & -1 \\ -2 & 1 & 2 \\ 0 & 1 & 3 \end{vmatrix} \underline{\underline{R_1(2)+R_2}}$$

$$- \begin{vmatrix} 1 & -1 & -1 \\ 0 & -1 & 0 \\ 0 & 1 & 3 \end{vmatrix} \underline{\underline{R_2(1)+R_3}} - \begin{vmatrix} 1 & -1 & -1 \\ 0 & -1 & 0 \\ 0 & 0 & 3 \end{vmatrix} = 3 \neq 0,$$

so D is invertible.

$|E| = 0$ since it contains a zero row, so E is not invertible.

$|F| = 1(2)(-6)(-1) = 12 \neq 0$, so F is invertible.

$$|A| = \begin{vmatrix} x & 0 & 0 & 0 \\ 2 & x & 2 & 0 \\ -2 & -3 & 1 & -2 \\ 1 & 0 & 0 & 1 \end{vmatrix} = \begin{vmatrix} x & 2 & -2 & 1 \\ 0 & x & -3 & 0 \\ 0 & 2 & 1 & 0 \\ 0 & 0 & -2 & 1 \end{vmatrix} \underline{\underline{R_3(-\frac{x}{2}) + R_2}}$$

$$\begin{vmatrix} x & 2 & -2 & 1 \\ 0 & 0 & -3-\frac{x}{2} & 0 \\ 0 & 2 & 1 & 0 \\ 0 & 0 & -2 & 1 \end{vmatrix} \underline{\underline{R_{2,3}}} - \begin{vmatrix} x & 2 & -2 & 1 \\ 0 & 2 & 1 & 0 \\ 0 & 0 & -3-\frac{x}{2} & 0 \\ 0 & 0 & -2 & 1 \end{vmatrix}$$

$$\underline{\underline{R_4(-\frac{3+\frac{x}{2}}{2}) + R_3}} - \begin{vmatrix} x & 2 & -2 & 1 \\ 0 & 2 & 1 & 0 \\ 0 & 0 & 0 & -\frac{3+\frac{x}{2}}{2} \\ 0 & 0 & -2 & 1 \end{vmatrix} \underline{\underline{R_{3,4}}} \begin{vmatrix} x & 2 & -2 & 1 \\ 0 & 2 & 1 & 0 \\ 0 & 0 & -2 & 1 \\ 0 & 0 & 0 & -\frac{6+x}{4} \end{vmatrix}$$

$$= x(6+x).$$

Let $|A| = 0$. Then $x(6+x) = 0$. So $x = 0$ or $x = -6$. Hence, when $x \neq 0$ and $x \neq -6$, $|A| \neq 0$ and A is invertible.

$$|B| = \begin{vmatrix} 2 & x & 4 \\ 1 & -2 & 4 \\ -2 & -1 & x \end{vmatrix} \underline{\underline{R_1(1) + R_3}} \begin{vmatrix} 2 & x & 4 \\ 1 & -2 & 4 \\ 0 & -1+x & 4+x \end{vmatrix} \underline{\underline{R_{1,2}}}$$

$$- \begin{vmatrix} 1 & -2 & 4 \\ 2 & x & 4 \\ 0 & -1+x & 4+x \end{vmatrix} \underline{\underline{R_1(-2) + R_2}} - \begin{vmatrix} 1 & -2 & 4 \\ 0 & x+4 & -4 \\ 0 & x-1 & 4+x \end{vmatrix}$$

$$= - \begin{vmatrix} 1 & 0 & 0 \\ -2 & x+4 & x-1 \\ 4 & -4 & 4+x \end{vmatrix} = - \begin{vmatrix} x+4 & x-1 \\ -4 & 4+x \end{vmatrix}$$

$$= -[(x+4)^2 + 4(x-1)] = -(x^2 + 12x + 12) = -[(x+6)^2 - 24].$$

Let $|B| = 0$. Then $(x+6)^2 = 24 = 2^2 6$. $x + 6 = \pm 2\sqrt{6}$ and $x = -6 \pm 2\sqrt{6}$.

Hence when $x \neq -6-2\sqrt{6}$ and $x \neq -6+2\sqrt{6}$, $|B| \neq 0$ and B is invertible.

4. $$|A+B| = \begin{vmatrix} 2 & 1 & 3 \\ 1 & 0 & 3 \\ 1 & 4 & 7 \end{vmatrix} + \begin{vmatrix} 2 & 1 & 3 \\ 1 & 0 & 3 \\ 0 & 1 & -1 \end{vmatrix} = \begin{vmatrix} 2 & 1 & 3 \\ 1 & 0 & 3 \\ 1+0 & 4+1 & 7-1 \end{vmatrix}$$
$$= \begin{vmatrix} 2 & 1 & 3 \\ 1 & 0 & 3 \\ 1 & 5 & 6 \end{vmatrix} \xlongequal[R_2(-1)+R_3]{R_2(-2)+R_1} \begin{vmatrix} 0 & 1 & -3 \\ 1 & 0 & 3 \\ 0 & 5 & 3 \end{vmatrix} \xlongequal{R_{1,2}}$$
$$- \begin{vmatrix} 1 & 0 & 3 \\ 0 & 1 & -3 \\ 0 & 5 & 3 \end{vmatrix} \xlongequal{R_2(-5)+R_3} - \begin{vmatrix} 1 & 0 & 3 \\ 0 & 1 & -3 \\ 0 & 0 & 18 \end{vmatrix} = -18.$$

5. $$|A+B| = \begin{vmatrix} 1 & 0 & 0 & 2 \\ -1 & 0 & 1 & 2 \\ 0 & 1 & 4 & 3 \\ -2 & 2 & 0 & 1 \end{vmatrix} + \begin{vmatrix} 1 & 0 & 0 & 2 \\ 3 & 0 & 0 & 1 \\ 0 & 1 & 4 & 3 \\ -2 & 2 & 0 & 1 \end{vmatrix}$$
$$= \begin{vmatrix} 1 & 0 & 0 & 2 \\ -1+3 & 0+0 & 1+0 & 2+1 \\ 0 & 1 & 4 & 3 \\ -2 & 2 & 0 & 1 \end{vmatrix} = \begin{vmatrix} 1 & 0 & 0 & 2 \\ 2 & 0 & 1 & 3 \\ 0 & 1 & 4 & 3 \\ -2 & 2 & 0 & 1 \end{vmatrix}$$
$$= \begin{vmatrix} 1 & 2 & 0 & -2 \\ 0 & 0 & 1 & 2 \\ 0 & 1 & 4 & 0 \\ 2 & 3 & 3 & 1 \end{vmatrix} \xlongequal{R_1(-2)+R_4} \begin{vmatrix} 1 & 2 & 0 & -2 \\ 0 & 0 & 1 & 2 \\ 0 & 1 & 4 & 0 \\ 0 & -1 & 3 & 5 \end{vmatrix} \xlongequal{R_3(1)+R_4}$$
$$\begin{vmatrix} 1 & 2 & 0 & -2 \\ 0 & 0 & 1 & 2 \\ 0 & 1 & 4 & 0 \\ 0 & 0 & 7 & 5 \end{vmatrix} \xlongequal{R_{2,3}} - \begin{vmatrix} 1 & 2 & 0 & -2 \\ 0 & 1 & 4 & 0 \\ 0 & 0 & 1 & 2 \\ 0 & 0 & 7 & 5 \end{vmatrix} \xlongequal{R_3(-7)+R_4}$$
$$- \begin{vmatrix} 1 & 2 & 0 & -2 \\ 0 & 1 & 4 & 0 \\ 0 & 0 & 1 & 2 \\ 0 & 0 & 0 & -9 \end{vmatrix} = 9.$$

□

A.4 Systems of linear equations

Section 4.1

1. Determine which of the following equations are linear.

   a) $2x+y=1.$ *b)* $0x+0y=3.$ *c)* $0x+0y=0.$ *d)* $x+\frac{2}{3}y=5.$
   e) $x^3+y^4=5.$ *f)* $xy=6.$ *g)* $3x^2+5y^2=1.$ *h)* $4xy+5y=3x.$
   i) $y=\cos x.$ *j)* $\frac{1}{\sqrt{3}}x+10^{\frac{1}{3}}y=4.$ *k)* $x^{\frac{1}{\sqrt{2}}}+6y^{\frac{1}{4}}=1.$
   l) $\frac{3}{\sqrt{4}}x=7^{\frac{1}{3}}x-2y+2.$

2. Consider the linear equation

$$2x+y=5.$$

   Verify whether the following points are the solutions of the above linear equation:

   $P_1(0,1)$, $P_2(1,0)$, $P_3(0,5)$ and $P_4(\frac{1}{2}x, 5-x)$ for each $x \in \mathbb{R}$.

3. Solve the linear equation

$$x+y=2$$

   and express the solution by a linear combination of two vectors.

4. Solve the linear equation $0x+0y=1$.

5. Verify whether the following points: $(1,-2)$, $(0,\frac{1}{2})$, $(1,1)$ are solutions of the linear equation

$$x-2y=1$$

   and solve this equation.

Solution.

1. *a)*, *b)*, *c)*, *d)*, *j)*, and *l)* are linear.
2. $P_3(0,5)$ and $P_4(\frac{1}{2}x, 5-x)$ are solutions of the equation.
3. Let $y=t$. Then $x=2-t$. Hence

$$\begin{pmatrix} x \\ y \end{pmatrix} = \begin{pmatrix} 2-t \\ t \end{pmatrix} = \begin{pmatrix} 2 \\ 0 \end{pmatrix} + \begin{pmatrix} -t \\ t \end{pmatrix} = \begin{pmatrix} 2 \\ 0 \end{pmatrix} + t\begin{pmatrix} -1 \\ 1 \end{pmatrix}.$$

4. The equation has no solutions.
5. Since

$$1-2(-2)=1+4=5\neq 1;\ 0-2(\frac{1}{2})=-1\neq 1; 1-2(1)=1-2=-1\neq 1,$$

so the three points are not solutions.

Let $y=t$. Then $x=1+2y=1+2t$. Hence $x=1+2t$ and $y=1$ is a solution of the equation. □

Section 4.2

1. Verify which of the following points:

$$P_1(6, -\frac{1}{2}), P_2(8,1), P_3(2,3), P_4(0,0)$$

are solutions of system of the linear equations

$$\begin{cases} x - 2y = 7 \\ x + 2y = 5 \end{cases}$$

2. Solve each of the following systems

$$(1) \begin{cases} x_1 - x_2 = 6 \\ x_1 + x_2 = 5 \end{cases} \qquad (2) \begin{cases} x_1 + 2x_2 = 1 \\ 2x_1 + x_2 = 2 \end{cases} \qquad (3) \begin{cases} x_1 + 2x_2 = 1 \\ x_1 + 2x_2 = 2 \end{cases}$$

Solution.

1. Since $6 - 2(-\frac{1}{2}) = 7$ and $6 + 2(-\frac{1}{2}) = 5$, $(6, -\frac{1}{2})$ is a solution of the system.

Since $8 - 2(1) = 6 \neq 7$, $(8,1)$ is not a solution of of the system.

Since $2 - 2(3) = 2 - 6 = -4 \neq 7$, $(2,3)$ is not a solution of of the system.

Since $0 - 2(0) = 0 \neq 7$, $(0,0)$ is not a solution of of the system.

2. (1). Adding the two equations implies $2x_1 = 11$ and $x_1 = \frac{11}{2}$. By the first equation, we have $x_2 = x_1 - 6 = \frac{11}{2} - 6 = -\frac{1}{2}$, so $(x_1, x_2) = (\frac{11}{2}, -\frac{1}{2})$ is a solution of (1)

(2). By the system (2), we have

$$(x_1 + 2x_2) - 2(2x_1 + x_2) = 1 - 2(2)$$

and $x_1 = 1$. By the second equation, $x_2 = 2 - 2x_1 = 2 - 2(1) = 0$. Hence $(x_1, x_2) = (1, 0)$ is a solution of (2).

(3). By the system (3), we have

$$(x_1 + 2x_2) - (x_1 + 2x_2) = 1 - 2.$$

This implies that $0 = -1$, a contradiction. Hence, (3) has no solutions. □

Section 4.3

1. Determine which of the following equations are linear.

a) $x + 2y + z = 6$. b) $2x - 6y + z = 1$. c) $-\sqrt{2}x + \frac{1}{6^{\frac{2}{3}}}y = 4 - 3z$.

d) $3x_1 + 2x_2 + 4x_3 + 5x_4 = 1$. e) $2xy + 3yz + 5z = 8$.

2. Find the coefficient matrix and the augmented matrix of each of the following systems of linear equations.

(1) $x+2y=6.$ (2) $\begin{cases} 2x_1 - x_2 = 6 \\ 4x_1 + x_2 = 3. \end{cases}$ (3) $\begin{cases} -x_1 + 2x_2 + 3x_3 = 4 \\ 3x_1 + 2x_2 - 3x_3 = 5 \\ 2x_1 + 3x_2 - x_3 = 1. \end{cases}$

(4) $\begin{cases} x_1 - 3x_3 - 4 = 0 \\ 2x_2 - 5x_3 - 8 = 0 \\ 3x_1 + 2x_2 - x_3 = 4. \end{cases}$

3. Find the system of linear equations corresponding to each of the following augmented matrices:

$$\left(\begin{array}{ccc|c} 1 & 2 & 3 & 1 \\ -1 & 1 & 0 & -2 \\ 0 & -1 & 1 & 0 \end{array}\right) \quad \left(\begin{array}{cc|c} 2 & 1 & 1 \\ 0 & -1 & -2 \\ 0 & 3 & -1 \end{array}\right).$$

4. Consider the following system of linear equations

a) $\begin{cases} -x_1 + x_2 + 2x_3 = 3 \\ 2x_1 + 6x_2 - 5x_3 = 2 \\ -3x_1 + 7x_2 - 5x_3 = -1. \end{cases}$ b) $\begin{cases} x_1 - x_2 + 4x_3 = 0 \\ -2x_1 + 4x_2 - 3x_3 = 0 \\ 3x_1 + 6x_2 - 8x_3 = 0. \end{cases}$

Write the above system into the form $A\overrightarrow{X} = \overrightarrow{b}$ and express $\overrightarrow{b}$ as a linear combination of the column vectors of the coefficient matrix A.

5. Consider the following system of linear equations

$$\begin{cases} x_1 + x_2 + 3x_3 = 2 \\ 2x_1 + 4x_2 - 3x_3 = 1 \\ 3x_1 + 6x_2 + 5x_3 = 0. \end{cases}$$

(1) Verify that $P_1(1, 3, -1)$ is not a solution of the system. $P_2(\frac{5}{6}, \frac{7}{6}, \frac{4}{3})$ is a solution of the system.

(2) Determine if the vector $\overrightarrow{b} = \begin{pmatrix} 3 \\ 2 \\ -1 \end{pmatrix}$ is a linear combination of the column vectors of the coefficient matrix of the system.

Solution.

1. a), b), c), d) are linear.
2. (1) $A = \begin{pmatrix} 1 & 2 \end{pmatrix}$, $(A|B) \left(\begin{array}{cc|c} 1 & 2 & 6 \end{array}\right)$.

(2) $A = \begin{pmatrix} 2 & -1 \\ 4 & 1 \end{pmatrix}$, $(A|B) = \left(\begin{array}{cc|c} 2 & -1 & 6 \\ 4 & 1 & 3 \end{array}\right)$

(3) $A = \begin{pmatrix} -1 & 2 & 3 \\ 3 & 2 & -3 \\ 2 & 3 & -1 \end{pmatrix}$, $(A|\overrightarrow{b}) = \left(\begin{array}{ccc|c} -1 & 2 & 3 & 4 \\ 3 & 2 & -3 & 5 \\ 2 & 3 & -1 & 1 \end{array}\right)$

(4) $A = \begin{pmatrix} 1 & 0 & -3 \\ 0 & 2 & -5 \\ 3 & 2 & -1 \end{pmatrix}$, $(A|\overrightarrow{b}) = \left(\begin{array}{ccc|c} 1 & 0 & -3 & 4 \\ 0 & 2 & -5 & 8 \\ 3 & 2 & -1 & 4 \end{array}\right)$

3.

$$\begin{cases} x_1 + 2x_2 + 3x_3 = 1 \\ -x_1 + x_2 = -2 \\ -x_2 + x_3 = 0 \end{cases} \qquad \begin{cases} 2x + y = 1 \\ -y = -2 \\ 3y = -1 \end{cases}$$

4 a) Let $A = \begin{pmatrix} -1 & 1 & 2 \\ 2 & 6 & -5 \\ -3 & 7 & -5 \end{pmatrix}$, $\overrightarrow{X} = \begin{pmatrix} x_1 \\ x_2 \\ x_3 \end{pmatrix}$ and $\overrightarrow{b} = \begin{pmatrix} 3 \\ 2 \\ -1 \end{pmatrix}$. Then $A\overrightarrow{X} = \overrightarrow{b}$.

Let $\overrightarrow{c_1} = \begin{pmatrix} -1 \\ 2 \\ -3 \end{pmatrix}$, $\overrightarrow{c_2} = \begin{pmatrix} 1 \\ 6 \\ 7 \end{pmatrix}$ and $\overrightarrow{c_3} = \begin{pmatrix} 2 \\ -5 \\ -5 \end{pmatrix}$. Then

$$x_1\overrightarrow{c_1} + x_2\overrightarrow{c_2} + x_3\overrightarrow{c_3} = \overrightarrow{b}.$$

b) Let $A = \begin{pmatrix} 1 & -1 & 4 \\ -2 & 4 & -3 \\ 3 & 6 & -8 \end{pmatrix}$, $\overrightarrow{X} = \begin{pmatrix} x_1 \\ x_2 \\ x_3 \end{pmatrix}$ and $\overrightarrow{b} = \begin{pmatrix} 0 \\ 0 \\ 0 \end{pmatrix}$. Then $A|\overrightarrow{X} = \overrightarrow{b}$.

Let $\overrightarrow{C_1} = \begin{pmatrix} 1 \\ -2 \\ 3 \end{pmatrix}$, $\overrightarrow{C_2} = \begin{pmatrix} -1 \\ 4 \\ 6 \end{pmatrix}$, $\overrightarrow{C_3} = \begin{pmatrix} 4 \\ -3 \\ -8 \end{pmatrix}$. Then

$$x_1\overrightarrow{C_1} + x_2\overrightarrow{C_2} + x_3\overrightarrow{C_3} = \overrightarrow{b}.$$

5. (1) Since $-1 + 3 + 2(-1) = 0 \neq 3$, $(1, 3, -1)$ is not a solution of the system. Since

$$\begin{cases} -\frac{5}{6} + \frac{7}{6} + 2(\frac{4}{3}) = \frac{-5+7+16}{6} = \frac{18}{6} = 3, \\ 2(\frac{5}{6}) + 6(\frac{7}{6}) - (5\frac{4}{3}) = \frac{5}{3} + 7 - \frac{20}{3} = 7 - 5 = 2, \\ -3(\frac{5}{6}) + 7(\frac{7}{6}) - 5(\frac{4}{3}) = -\frac{15}{6} + \frac{49}{6} - \frac{40}{6} = -1, \end{cases}$$

$(\frac{5}{6}, \frac{7}{6}, \frac{4}{3})$ is a solution of the system.

(2) Since $(\frac{5}{6}, \frac{7}{6}, \frac{4}{3})$ is a solution of the system, $\overrightarrow{b} = \begin{pmatrix} 3 \\ 2 \\ -1 \end{pmatrix}$ is a linear combination of the column vectors of the coefficient matrix. That is, let $\overrightarrow{c_1} = \begin{pmatrix} -1 \\ 2 \\ -3 \end{pmatrix}$, $\overrightarrow{c_2} = \begin{pmatrix} 1 \\ 6 \\ 7 \end{pmatrix}$ and $\overrightarrow{c_3} = \begin{pmatrix} 2 \\ -5 \\ -5 \end{pmatrix}$. Then

$$\frac{5}{6}\overrightarrow{c_1} + \frac{7}{6}\overrightarrow{c_2} + \frac{4}{3}\overrightarrow{c_3} = \overrightarrow{b}.$$

□

Section 4.4

1. Consider the system of the following linear equations

$$\begin{cases} x_1 + x_2 - 3x_3 = 2 \\ -x_2 - 6x_3 = 4 \\ -x_3 = 1. \end{cases}$$

(a) Find the coefficient matrix and the augmented matrix of the system.

(b) Find the basic variables and free variables of the system.

(c) Solve the system by using back-substitution.

(d) Determine if the vector $\overrightarrow{b} = \begin{pmatrix} 2 \\ 4 \\ 1 \end{pmatrix}$ is a linear combination of the column vectors of the coefficient matrix of the system.

2. Consider the following system

$$\begin{cases} x_1 - 2x_2 + 3x_3 + x_4 = -1 \\ x_2 - x_3 + x_4 = 4 \\ -x_4 = 2. \end{cases}$$

(a) Find the coefficient matrix and the augmented matrix of the system.

(b) Find the basic variables and free variables of the system.

(c) Solve the system by using back-substitution.

(d) If the system is consistent, then express its solution as a linear combination.

(e) Determine if the vector $\overrightarrow{b} = \begin{pmatrix} -1 \\ 4 \\ 2 \end{pmatrix}$ is a linear combination of the column vectors of the coefficient matrix of the system.

3. Consider the following system

$$\begin{cases} -x_1 - 2x_2 - 2x_3 + x_4 - 3x_5 = 4 \\ -2x_3 + 3x_4 = 1 \\ -4x_4 + 2x_5 = 5. \end{cases}$$

(*a*) Find the coefficient matrix and the augmented matrix of the system.

(*b*) Find the basic variables and free variables of the system.

(*c*) Solve the system.

(*d*) Express the solutions as a linear combination.

(*e*) Determine if the vector $\overrightarrow{b} = \begin{pmatrix} 4 \\ 1 \\ 5 \end{pmatrix}$ is a linear combination of the column vectors of the coefficient matrix of the system.

4. Consider the following system

$$\begin{cases} -3x_1 - 2x_2 - 2x_3 = 2 \\ x_2 - x_3 = 2 \\ 0x_3 = 4. \end{cases}$$

(*a*) Find the coefficient matrix and the augmented matrix of the system.

(*b*) Find the basic variables and free variables of the system.

(*c*) Solve the system.

Solution.

1. (*a*) $A = \begin{pmatrix} 1 & 1 & -3 \\ 0 & -1 & -6 \\ 0 & 0 & -1 \end{pmatrix}$, $(A|\overrightarrow{b}) = \left(\begin{array}{ccc|c} 1 & 1 & -3 & 2 \\ 0 & -1 & -6 & 4 \\ 0 & 0 & -1 & 1 \end{array}\right)$

(*b*) Since A is a row echelon matrix, the basic variables are x_1, x_2, x_3. There are no free variables.

(*c*) By the third equation, $x_3 = -1$. By the second equation, $x_2 = -6x_3 - 4 = -6(-1) - 4 = 2$ and by the first equation,

$$x_1 = 2 - x_2 + 3x_3 = 2 - 2 + 3(-1) = -3.$$

Hence, $(x_1, x_2, x_3) = (-3, 2, -1)$ is a solution.

(*d*) Since the system has a solution $(-3, 2, -1)$, $\overrightarrow{b} = \begin{pmatrix} 2 \\ 4 \\ 1 \end{pmatrix}$ is a linear combination of the column vectors of the coefficient matrix.

2. (a) $A = \begin{pmatrix} 1 & -2 & 3 & 1 \\ 0 & 1 & -1 & 1 \\ 0 & 0 & 0 & -1 \end{pmatrix}$ and

$$(A|\overrightarrow{b}) = \left(\begin{array}{cccc|c} 1 & -2 & 3 & 1 & -1 \\ 0 & 1 & -1 & 1 & 4 \\ 0 & 0 & 0 & -1 & 2 \end{array}\right).$$

(b) x_1, x_2, x_4 are basic variables and x_3 is a free variable.
(c) By the third equation, $x_4 = -2$; Let $x_3 = t$. Then by the second equation, $x_2 = 4 - x_4 + x_3 = 4 + 2 + x_3 = 6 + t$. By the first equation,

$$x_1 = -1 + 2x_2 - 3x_3 - x_4 = -1 + 2(6+t) - 3t - (-2) = 13 - t.$$

(d) $\begin{pmatrix} x_1 \\ x_2 \\ x_3 \\ x_4 \end{pmatrix} = \begin{pmatrix} 13-t \\ 6+t \\ t \\ -2 \end{pmatrix} = \begin{pmatrix} 13 \\ 6 \\ 0 \\ -2 \end{pmatrix} + \begin{pmatrix} -t \\ t \\ t \\ 0 \end{pmatrix} = \begin{pmatrix} 13 \\ 6 \\ 0 \\ -2 \end{pmatrix} + t \begin{pmatrix} -1 \\ 1 \\ 1 \\ 0 \end{pmatrix}.$

(e) Yes, since the system is consistent.

3. (a) $A = \begin{pmatrix} -1 & -2 & -2 & 1 & -3 \\ 0 & 0 & -2 & 3 & 0 \\ 0 & 0 & 0 & -4 & 2 \end{pmatrix}$ and

$$(A|\overrightarrow{b}) = \left(\begin{array}{ccccc|c} -1 & -2 & -2 & 1 & -3 & 4 \\ 0 & 0 & -2 & 3 & 0 & 1 \\ 0 & 0 & 0 & -4 & 2 & 5 \end{array}\right).$$

(b) x_1, x_3, x_4 are basic variables and x_2 and x_5 are free variables.
(c) Let $x_2 = s$ and $x_5 = t$. By the third equation,

$$-4x_4 = 5 - 2x_5 \quad \text{and } x_4 = -\tfrac{5}{4} + \tfrac{1}{2}t.$$

By the second equation, we have

$$-2x_3 = 1 - 3x_4 = 1 - 3(-\frac{5}{4} + \frac{1}{2}t) = 1 + \frac{15}{4} - \frac{3}{2}t = \frac{19}{4} - \frac{3}{2}t$$

and $x_3 = -\frac{19}{8} + \frac{3}{4}t$. By the first equation,

$$\begin{aligned} x_1 &= -4 - 2x_2 - 2x_3 + x_4 - 3x_5 = -4 - 2s - 2(-\frac{19}{8} + \frac{3}{4}t) + (-\frac{5}{4} + \frac{1}{2}t) - 3t \\ &= -4 - 2s + \frac{19}{4} - \frac{3}{2}t - \frac{5}{4} + \frac{1}{2}t - 3t = \frac{-16 + 19 - 5}{4} - 2s + (\frac{-3 + 1 - 6}{2})t \\ &= -\frac{1}{2} - 2s - 4t. \end{aligned}$$

(*d*)

$$\begin{pmatrix} x_1 \\ x_2 \\ x_3 \\ x_4 \\ x_5 \end{pmatrix} = \begin{pmatrix} -\frac{1}{2} - 2s - 4t \\ s \\ -\frac{19}{8} + \frac{3}{4}t \\ -\frac{5}{4} + \frac{1}{2}t \\ t \end{pmatrix} = \begin{pmatrix} -\frac{1}{2} \\ 0 \\ -\frac{19}{8} \\ -\frac{5}{4} \\ 0 \end{pmatrix} + \begin{pmatrix} -2s \\ s \\ 0 \\ 0 \\ 0 \end{pmatrix} + \begin{pmatrix} -4t \\ 0 \\ \frac{3}{4}t \\ \frac{1}{2}t \\ t \end{pmatrix}$$

$$= \begin{pmatrix} -\frac{1}{2} \\ 0 \\ -\frac{19}{8} \\ -\frac{5}{4} \\ 0 \end{pmatrix} + s\begin{pmatrix} -2 \\ 1 \\ 0 \\ 0 \\ 0 \end{pmatrix} + t\begin{pmatrix} -4 \\ 0 \\ \frac{3}{4} \\ \frac{1}{2} \\ 1 \end{pmatrix}.$$

(*e*) Yes, since the system is consistent.

4. (*a*) $A = \begin{pmatrix} -3 & -2 & -2 \\ 0 & 1 & -1 \\ 0 & 0 & 0 \end{pmatrix}$ and $(A|\overrightarrow{b}) = \left(\begin{array}{ccc|c} -3 & -2 & -2 & 2 \\ 0 & 1 & -1 & 2 \\ 0 & 0 & 0 & 4 \end{array}\right)$

(*b*) x_1, x_2 are basic variables and x_3 is a free variable.

(*c*) The system has no solutions. □

Section 4.5

1. Use Gaussian elimination to solve the following systems and express the solutions as linear combinations if they have infinitely many solutions.

a) $\begin{cases} x_1 - 2x_2 + 3x_3 = 6 \\ 2x_1 + x_2 + 4x_3 = 5 \\ -3x_1 + x_2 - 2x_3 = 3. \end{cases}$ b) $\begin{cases} x_1 + x_2 - 2x_3 = 3 \\ 2x_1 + 3x_2 - x_3 = -4 \\ -2x_1 - 3x_2 - x_3 = 4 \end{cases}$

c) $\begin{cases} x_1 + 2x_2 + 3x_3 = 4 \\ 4x_1 + 7x_2 + 6x_3 = 17 \\ 2x_1 + 5x_2 + 12x_3 = 7. \end{cases}$

Solution. 1. *a*)

$$(A|\overrightarrow{b}) = \left(\begin{array}{ccc|c} 1 & -2 & 3 & 6 \\ 2 & 1 & 4 & 5 \\ -3 & 1 & -2 & 3 \end{array}\right) \xrightarrow[R_1(3)+R_3]{R_1(-2)+R_2} \left(\begin{array}{ccc|c} 1 & -2 & 3 & 6 \\ 0 & 5 & -2 & -7 \\ 0 & -5 & 7 & 21 \end{array}\right)$$

$$\xrightarrow{R_2(1)+R_3} \left(\begin{array}{ccc|c} 1 & -2 & 3 & 6 \\ 0 & 5 & -2 & -7 \\ 0 & 0 & 5 & 14 \end{array}\right).$$

Hence, we obtain

$$\begin{cases} x_1 - 2x_2 + 3x_3 = 6 & (1) \\ 5x_2 - 2x_3 = -7 & (2) \\ 5x_3 = 14. & (3) \end{cases}$$

By (3), $x_3 = \frac{14}{5}$. By (2),

$$5x_2 = -7 + 2x_3 = -7 + 2(\frac{14}{5}) = -7 + \frac{28}{5} = -\frac{7}{5},$$

and $x_2 = -\frac{7}{25}$. By (1), we have

$$x_1 = 6 + 2x_2 - 3x_3 = 6 + 2(-\frac{7}{25}) - 3(\frac{14}{5}) = 6 - \frac{14}{25} - \frac{210}{25} = -\frac{74}{25}.$$

$$b) \quad (A|\vec{b}) = \left(\begin{array}{ccc|c} 1 & 1 & -2 & 3 \\ 2 & 3 & -1 & -4 \\ -2 & -3 & -1 & 4 \end{array}\right) \xrightarrow[R_1(2)+R_3]{R_1(-2)+R_2}$$

$$\left(\begin{array}{ccc|c} 1 & 1 & -2 & 3 \\ 0 & 1 & 3 & -10 \\ 0 & -1 & -5 & 10 \end{array}\right) \xrightarrow{R_2(1)+R_3} \left(\begin{array}{ccc|c} 1 & 1 & -2 & 3 \\ 0 & 1 & 3 & -10 \\ 0 & 0 & -2 & 0 \end{array}\right).$$

Hence, we obtain

$$\begin{cases} x_1 + x_2 - 2x_3 = 3 & (1) \\ x_2 + 3x_3 = 10 & (2) \\ -2x_3 = 0. & (3) \end{cases}$$

By (3), $x_3 = 0$. By (2), $x_2 = -10 - 3x_3 = -10$. By (1),

$$x_1 = 3 - x_2 + 2x_3 = 3 - (-10) = 13.$$

Hence, $(x_1, x_2, x_3) = (13, -10, 0)$ is a solution.

$$(c) \quad (A|\vec{b}) = \left(\begin{array}{ccc|c} 1 & 2 & 3 & 4 \\ 4 & 7 & 6 & 17 \\ 2 & 5 & 12 & 7 \end{array}\right) \xrightarrow[R_1(-2)+R_3]{R_1(-4)+R_2}$$

$$\left(\begin{array}{ccc|c} 1 & 2 & 3 & 4 \\ 0 & -1 & -6 & 1 \\ 0 & 1 & 6 & -1 \end{array}\right) \xrightarrow{R_2(1)+R_3} \left(\begin{array}{ccc|c} 1 & 2 & 3 & 4 \\ 0 & -1 & -6 & 1 \\ 0 & 0 & 0 & 0 \end{array}\right).$$

Hence, we have

$$\begin{cases} x_1 + 2x_2 + 3x_3 = 4 & (1) \\ -x_2 - 6x_3 = 1 & (2) \end{cases}$$

Let $x_3 = t$. By (2), $x_2 = -1 - 6x_3 = -1 - 6t$. By (1),

$$x_1 = 4 - 2x_2 - 3x_3 = 4 - 2(-1-6t) - 3t = 4 + 2 + 12t - 3t = 6 + 9t.$$

Hence,

$$\begin{pmatrix} x_1 \\ x_2 \\ x_3 \end{pmatrix} = \begin{pmatrix} 6+9t \\ -1-6t \\ t \end{pmatrix} = \begin{pmatrix} 6 \\ -1 \\ 0 \end{pmatrix} + \begin{pmatrix} 9t \\ -6t \\ t \end{pmatrix} = \begin{pmatrix} 6 \\ -1 \\ 0 \end{pmatrix} + t \begin{pmatrix} 9 \\ -6 \\ 1 \end{pmatrix}.$$

□

Section 4.6

1. Use Theorem 4.6.1 to determine whether the following systems have a unique solution, infinitely many solutions or no solutions.

a) $\begin{cases} x_1 - 2x_2 + 3x_3 = 6 \\ 2x_1 + x_2 + 4x_3 = 5 \\ -3x_1 + x_2 - 2x_3 = 3. \end{cases}$ b) $\begin{cases} x_1 + x_2 - 2x_3 = 3 \\ 2x_1 + 3x_2 - x_3 = -4 \\ -2x_1 - 3x_2 - x_3 = 4 \end{cases}$

c) $\begin{cases} x_1 + 2x_2 + 3x_3 = 4 \\ 4x_1 + 7x_2 + 6x_3 = 17 \\ 2x_1 + 5x_2 + 12x_3 = 7. \end{cases}$ d) $\begin{cases} x_1 - 2x_2 - x_3 + 2x_4 = 0 \\ -x_2 - 2x_3 + x_4 = 0. \end{cases}$

e) $\begin{cases} x_1 - x_2 + 3x_3 = 2 \\ -2x_1 + 2x_2 - 6x_3 = 5 \\ 2x_1 - 2x_2 + 6x_3 = 4. \end{cases}$

Solution. a)

$$(A|\vec{b}) = \left(\begin{array}{ccc|c} 1 & -2 & 3 & 6 \\ 2 & 1 & 4 & 5 \\ -3 & 1 & -2 & 3 \end{array}\right) \xrightarrow[R_1(3)+R_3]{R_1(-2)+R_2} \left(\begin{array}{ccc|c} 1 & -2 & 3 & 6 \\ 0 & 5 & -2 & -7 \\ 0 & -5 & 7 & 21 \end{array}\right)$$

$$\xrightarrow{R_2(1)+R_3} \left(\begin{array}{ccc|c} 1 & -2 & 3 & 6 \\ 0 & 5 & -2 & -7 \\ 0 & 0 & 5 & 14 \end{array}\right).$$

Hence, $r(A) = r(A|\overrightarrow{b}) = 3$. By Theorem 4.6.1 (1), the system has a unique solution.

$$b)\quad (A|\overrightarrow{b}) = \left(\begin{array}{ccc|c} 1 & 1 & -2 & 3 \\ 2 & 3 & -1 & -4 \\ -2 & -3 & -1 & 4 \end{array}\right) \xrightarrow[R_1(2)+R_3]{R_1(-2)+R_2} \left(\begin{array}{ccc|c} 1 & 1 & -2 & 6 \\ 0 & 1 & 3 & -10 \\ 0 & -1 & -5 & -2 \end{array}\right) \xrightarrow{R_2(1)+R_3} \left(\begin{array}{ccc|c} 1 & 1 & -2 & 6 \\ 0 & 1 & 3 & -10 \\ 0 & 0 & -2 & -12 \end{array}\right).$$

Hence, $r(A) = r(A|\overrightarrow{b}) = 3$. By Theorem 4.6.1 (1), the system has a unique solution.

$$c)\quad (A|\overrightarrow{b}) = \left(\begin{array}{ccc|c} 1 & 2 & 3 & 4 \\ 4 & 7 & 6 & 17 \\ 2 & 5 & 12 & 7 \end{array}\right) \xrightarrow[R_1(-2)+R_3]{R_1(-4)+R_2} \left(\begin{array}{ccc|c} 1 & 2 & 3 & 4 \\ 0 & -1 & -6 & 1 \\ 0 & 1 & 6 & -1 \end{array}\right) \xrightarrow{R_2(1)+R_3} \left(\begin{array}{ccc|c} 1 & 2 & 3 & 4 \\ 0 & -1 & 6 & 1 \\ 0 & 0 & 0 & 0 \end{array}\right).$$

Hence, $r(A) = r(A|\overrightarrow{b}) = 2 < 3$, by Theorem 4.6.1 (2), the system has infinitely many solutions.

(*d*) Since $n = 4$ and $m = 2$, by Corollary 4.6.5, the system has infinitely many solutions.

$$e)\quad (A|\overrightarrow{b}) = \left(\begin{array}{ccc|c} 1 & -1 & 3 & 2 \\ -2 & 2 & -6 & 5 \\ 2 & -2 & 6 & 4 \end{array}\right) \xrightarrow[R_1(-2)+R_3]{R_1(2)+R_2} \left(\begin{array}{ccc|c} 1 & -2 & 3 & 2 \\ 0 & 0 & 0 & 9 \\ 0 & 0 & 0 & 0 \end{array}\right).$$

Hence, $r(A) = 1$ and $r(A|\overrightarrow{b}) = 2$. By Theorem 4.6.1 (3), the system has no solutions. □

Section 4.7

1. Find conditions on b_1, b_2, b_3 such that each of the systems is consistent.

$$a)\ \begin{cases} x_1 - x_2 + 2x_3 = b_1 \\ 2x_1 + 4x_2 - 4x_3 = b_2 \\ -3x_1 + 4x_3 = b_3. \end{cases} \qquad b)\ \begin{cases} x_1 - 2x_2 + 2x_3 = b_1 \\ -3x_1 + 3x_2 + x_3 = b_2 \\ 4x_1 - 6x_2 + 3x_3 = b_3. \end{cases}$$

2. Consider the following system

$$\begin{cases} 2x_1 - 4x_2 = b_1 \\ x_1 - 2x_2 = b_2. \end{cases}$$

(1) Find b_1 and b_2 such that the system is consistent.

(2) Find b_1 and b_2 such that the system is inconsistent.

(3) If $b_1 = 4$ and $b_2 = 1$, determine whether the system is inconsistent.

3. Consider the system of linear equations

$$\begin{cases} x_1 - x_2 + x_3 = b_1 \\ -2x_1 - x_3 = b_2 \\ 2x_1 - 10x_2 + 6x_3 = b_3. \end{cases}$$

(1) Find conditions on b_1, b_2, b_3 such that the system is consistent.

(2) Find conditions on b_1, b_2, b_3 such that the system is inconsistent.

(3) Find a vector (b_1, b_2, b_3) such that the system is inconsistent.

Solution. 1. *a*)

$$(A|\overrightarrow{b}) = \left(\begin{array}{ccc|c} 1 & -1 & 2 & b_1 \\ 2 & 4 & -4 & b_2 \\ -3 & 0 & 4 & b_3 \end{array}\right) \xrightarrow[R_1(3)+R_3]{R_1(-2)+R_2} \left(\begin{array}{ccc|c} 1 & -1 & 2 & b_1 \\ 0 & 6 & -8 & b_2 - 2b_1 \\ 0 & -3 & 10 & b_3 + 3b_1 \end{array}\right)$$

$$\xrightarrow{R_2(\frac{1}{2})+R_3} \left(\begin{array}{ccc|c} 1 & -1 & 2 & b_1 \\ 0 & 6 & -8 & b_2 - 2b_1 \\ 0 & 0 & 6 & b_3 + 3b_1 + \frac{1}{2}(b_2 - 2b_1) \end{array}\right).$$

Since $r(A) = 3$, by Theorem 4.6.1 (1), the system has a unique solutions for any b_1, b_2, b_3.

$$b)\quad (A|\overrightarrow{b}) = \left(\begin{array}{ccc|c} 1 & -2 & 2 & b_1 \\ -3 & 3 & 1 & b_2 \\ 4 & -6 & 3 & b_3 \end{array}\right) \xrightarrow[R_1(-4)+R_3]{R_1(3)+R_2}$$

$$\left(\begin{array}{ccc|c} 1 & -2 & 2 & b_1 \\ 0 & -3 & 7 & b_2 + 3b_1 \\ 0 & 2 & -5 & b_3 - 4b_1 \end{array}\right) \xrightarrow{R_2(\frac{2}{3})+R_3}$$

$$\left(\begin{array}{ccc|c} 1 & -2 & 2 & b_1 \\ 0 & -3 & 7 & b_2 + 3b_1 \\ 0 & 0 & -\frac{1}{3} & b_3 - 4b_1 + \frac{2}{3}(b_2 + 3b_1) \end{array}\right).$$

Since $r(A) = 3$, by Theorem 4.6.1 (1), the system has a unique solutions for any b_1, b_2, b_3.

$$2.\quad (A|\overrightarrow{b}) = \left(\begin{array}{cc|c} 2 & -4 & b_1 \\ 1 & -2 & b_2 \end{array}\right) \xrightarrow{R_{1,2}} \left(\begin{array}{cc|c} 1 & -2 & b_2 \\ 2 & -4 & b_1 \end{array}\right) \xrightarrow{R_1(-2)+R_2}$$

$$\left(\begin{array}{cc|c} 1 & -2 & b_2 \\ 0 & 0 & b_1 - 2b_2 \end{array}\right).$$

So if $b_1 - 2b_2 = 0$, then $r(A) = r(A|\overrightarrow{b}) = 1 < 2$. By Theorem 4.6.1 (2), the system has infinite many solutions. If $b_1 - 2b_2 \neq 0$, then $r(A) = 1 < 2$ and $r(A|\overrightarrow{b}) = 2$. By Theorem 4.6.1 (3), the system has no solution.

(1) If $b_1 = 2b_2$, then the system is consistent

(2) If $b_1 \neq 2b_2$, then the system is inconsistent

(3) Let $\overrightarrow{b} = \begin{pmatrix} b_1 \\ b_2 \end{pmatrix} = \begin{pmatrix} 4 \\ 1 \end{pmatrix}$. Then $b_1 = 4$, $b_2 = 1$. Since $b_1 \neq 2b_2$, the system with $\overrightarrow{b} = \begin{pmatrix} 4 \\ 1 \end{pmatrix}$ is inconsistent.

3. $$(A|\overrightarrow{b}) = \left(\begin{array}{ccc|c} 1 & -1 & 1 & b_1 \\ -2 & 0 & -1 & b_2 \\ 2 & -10 & 6 & b_3 \end{array}\right) \xrightarrow[R_1(-2)+R_3]{R_1(2)+R_2} \left(\begin{array}{ccc|c} 1 & -1 & 1 & b_1 \\ 0 & -2 & 1 & b_2 + 2b_1 \\ 0 & -8 & 4 & b_3 - 2b_1 \end{array}\right)$$

$$\xrightarrow{R_2(-4)+R_3} \left(\begin{array}{ccc|c} 1 & -1 & 1 & b_1 \\ 0 & -2 & 1 & b_2 + 2b_1 \\ 0 & 0 & 0 & b_3 - 2b_1 - 4(b_2 + 2b_1) \end{array}\right)$$

$$= \left(\begin{array}{ccc|c} 1 & -1 & 1 & b_1 \\ 0 & -2 & 1 & b_2 + 2b_1 \\ 0 & 0 & 0 & b_3 - 4b_2 - 10b_1 \end{array}\right).$$

(1) If $b_3 - 4b_2 - 10b_1 = 0$, then the system is consistent.

(2) If $b_3 - 4b_2 - 10b_1 \neq 0$, then the system is inconsistent.

(3) To find a vector $b = \begin{pmatrix} b_1 \\ b_2 \\ b_3 \end{pmatrix}$ such that the system is inconsistent, we need to choose b_1, b_2, b_3 such that $b_3 - 4b_2 - 10b_1 \neq 0$. Hence, let $b_1 = b_2 = 0$ and $b_3 = 1$. Then $b_3 - 4b_2 - 10b_1 = 1 \neq 0$. Hence, the system with $b = \begin{pmatrix} 0 \\ 0 \\ 1 \end{pmatrix}$ is inconsistent. □

Section 4.8

1. Solve each of the following systems using Gauss-Jordan elimination.

$$a) \begin{cases} x_1 + 3x_2 - x_3 = 2 \\ 4x_1 - 6x_2 + 6x_3 = 14 \\ -3x_1 - x_2 - 2x_3 = 3. \end{cases} \qquad b) \begin{cases} x_1 - 2x_2 + 2x_3 = 2 \\ -2x_1 + 3x_2 - 4x_3 = -2 \\ -3x_1 + 4x_2 + 6x_3 = 0. \end{cases}$$

$$c) \begin{cases} x_2 - x_3 = 1 \\ -x_1 + 3x_2 - x_3 = -2 \\ 3x_1 + 3x_2 - 2x_3 = 4. \end{cases} \qquad d) \begin{cases} x_1 + 4x_2 + 6x_3 = 0 \\ 2x_1 + 6x_2 + 3x_3 = 0 \\ -3x_1 + x_2 - 2x_3 = 0. \end{cases}$$

$$e)\ \begin{cases} -x_1 + 2x_2 - x_3 = 0 \\ 3x_1 - 3x_2 - 9x_3 = 0 \\ -2x_1 - x_2 + 12x_3 = 0. \end{cases}$$

Solution.

1. $a)$ $(A|\overrightarrow{b}) = \left(\begin{array}{ccc|c} 1 & 3 & -1 & 2 \\ 4 & -6 & 6 & 14 \\ -3 & -1 & -2 & 3 \end{array}\right) \xrightarrow[R_1(3)+R_3]{R_1(-4)+R_2}$

$$\left(\begin{array}{ccc|c} 1 & 3 & -1 & 2 \\ 0 & -18 & 10 & 6 \\ 0 & 8 & -5 & 9 \end{array}\right) \xrightarrow{R_2(-\frac{1}{2})} \left(\begin{array}{ccc|c} 1 & 3 & -1 & 2 \\ 0 & 9 & -5 & -3 \\ 0 & 8 & -5 & 9 \end{array}\right)$$

$$\xrightarrow{R_3(-1)+R_2} \left(\begin{array}{ccc|c} 1 & 3 & -1 & 2 \\ 0 & 1 & 0 & -12 \\ 0 & 8 & -5 & 9 \end{array}\right) \xrightarrow{R_2(-8)+R_3}$$

$$\left(\begin{array}{ccc|c} 1 & 3 & -1 & 2 \\ 0 & 1 & 0 & -12 \\ 0 & 0 & -5 & 105 \end{array}\right) \xrightarrow{R_3(-\frac{1}{5})} \left(\begin{array}{ccc|c} 1 & 3 & -1 & 2 \\ 0 & 1 & 0 & -12 \\ 0 & 0 & 1 & -21 \end{array}\right)$$

$$\xrightarrow{R_3(1)+R_1} \left(\begin{array}{ccc|c} 1 & 3 & 0 & -19 \\ 0 & 1 & 0 & -12 \\ 0 & 0 & 1 & -21 \end{array}\right) \xrightarrow{R_2(-3)+R_1}$$

$$\left(\begin{array}{ccc|c} 1 & 0 & 0 & 17 \\ 0 & 1 & 0 & -12 \\ 0 & 0 & 1 & -21 \end{array}\right).$$

Hence, $x_1 = 17$, $x_2 = -12$, $x_3 = -21$ is a solution of the system.

$b)$ $(A|\overrightarrow{b}) = \left(\begin{array}{ccc|c} 1 & -2 & 2 & 2 \\ -2 & 3 & -4 & -2 \\ -3 & 4 & 6 & 0 \end{array}\right) \xrightarrow[R_1(3)+R_3]{R_1(2)+R_2} \left(\begin{array}{ccc|c} 1 & -2 & 2 & 2 \\ 0 & -1 & 0 & 2 \\ 0 & -2 & 12 & 6 \end{array}\right)$

$$\xrightarrow[R_3(-\frac{1}{2})]{R_2(-1)} \left(\begin{array}{ccc|c} 1 & -2 & 2 & 2 \\ 0 & 1 & 0 & -2 \\ 0 & 1 & -6 & -3 \end{array}\right) \xrightarrow{R_2(-1)+R_3} \left(\begin{array}{ccc|c} 1 & -2 & 2 & 2 \\ 0 & 1 & 0 & -2 \\ 0 & 0 & -6 & -1 \end{array}\right)$$

$$\xrightarrow{R_3(-\frac{1}{6})} \left(\begin{array}{ccc|c} 1 & -2 & 2 & 2 \\ 0 & 1 & 0 & -2 \\ 0 & 0 & 1 & \frac{1}{6} \end{array}\right) \xrightarrow{R_3(-2)+R_1} \left(\begin{array}{ccc|c} 1 & -2 & 0 & \frac{5}{3} \\ 0 & 1 & 0 & -2 \\ 0 & 0 & 1 & \frac{1}{6} \end{array}\right)$$

$$\xrightarrow{R_2(2)+R_1} \left(\begin{array}{ccc|c} 1 & 0 & 0 & -\frac{7}{3} \\ 0 & 1 & 0 & -2 \\ 0 & 0 & 1 & \frac{1}{6} \end{array}\right).$$

Hence, $x_1 = -\frac{7}{3}$, $x_2 = -2$, $x_3 = \frac{1}{6}$ is a solution of the system.

c) $(A|\overrightarrow{b}) = \left(\begin{array}{ccc|c} 0 & 1 & -1 & 1 \\ -1 & 3 & -1 & -2 \\ 3 & 3 & -2 & 4 \end{array}\right) \xrightarrow{R_{1,2}} \left(\begin{array}{ccc|c} -1 & 3 & -1 & -2 \\ 0 & 1 & -1 & 1 \\ 3 & 3 & -2 & 4 \end{array}\right)$

$$\xrightarrow{R_1(3)+R_3} \left(\begin{array}{ccc|c} -1 & 3 & -1 & -2 \\ 0 & 1 & -1 & 1 \\ 0 & 12 & -5 & -2 \end{array}\right) \xrightarrow[R_2(-12)+R_3]{R_1(-1)}$$

$$\left(\begin{array}{ccc|c} 1 & -3 & 1 & 2 \\ 0 & 1 & -1 & 1 \\ 0 & 0 & 7 & -14 \end{array}\right) \xrightarrow{R_3(\frac{1}{7})} \left(\begin{array}{ccc|c} 1 & -3 & 1 & 2 \\ 0 & 1 & -1 & 1 \\ 0 & 0 & 1 & -2 \end{array}\right)$$

$$\xrightarrow[R_3(1)+R_2]{R_3(-1)+R_1} \left(\begin{array}{ccc|c} 1 & -3 & 0 & 4 \\ 0 & 1 & 0 & -1 \\ 0 & 0 & 1 & -2 \end{array}\right) \xrightarrow{R_2(3)+R_1} \left(\begin{array}{ccc|c} 1 & 0 & 0 & 1 \\ 0 & 1 & 0 & -1 \\ 0 & 0 & 1 & -2 \end{array}\right).$$

Hence, $x_1 = 1$, $x_2 = -1$, $x_3 = -2$ is a solution of the system.

d) $(A|\overrightarrow{0}) = \left(\begin{array}{ccc|c} 1 & 4 & 6 & 0 \\ 2 & 6 & 3 & 0 \\ -3 & 1 & -2 & 0 \end{array}\right) \xrightarrow[R_1(3)+R_3]{R_1(-2)+R_2} \left(\begin{array}{ccc|c} 1 & 4 & 6 & 0 \\ 0 & -2 & -9 & 0 \\ 0 & 13 & 16 & 0 \end{array}\right)$

$$\xrightarrow{R_3(\frac{1}{13})} \left(\begin{array}{ccc|c} 1 & 4 & 6 & 0 \\ 0 & -2 & -9 & 0 \\ 0 & 1 & 2 & 0 \end{array}\right) \xrightarrow[R_3(2)+R_2]{R_3(-4)+R_1}$$

$$\left(\begin{array}{ccc|c} 1 & 0 & -2 & 0 \\ 0 & 0 & -5 & 0 \\ 0 & 1 & 2 & 0 \end{array}\right) \xrightarrow{R_{2,3}} \left(\begin{array}{ccc|c} 1 & 0 & -2 & 0 \\ 0 & 1 & 2 & 0 \\ 0 & 0 & -5 & 0 \end{array}\right)$$

$$\xrightarrow{R_3(-\frac{1}{5})} \left(\begin{array}{ccc|c} 1 & 0 & -2 & 0 \\ 0 & 1 & 2 & 0 \\ 0 & 0 & 1 & 0 \end{array}\right) \xrightarrow[R_3(-2)+R_2]{R_3(2)+R_1} \left(\begin{array}{ccc|c} 1 & 0 & 0 & 0 \\ 0 & 1 & 0 & 0 \\ 0 & 0 & 1 & 0 \end{array}\right).$$

Hence, $x_1 = x_2 = x_3 = 0$ is a solution of the system.

e) $(A|\overrightarrow{0}) = \left(\begin{array}{ccc|c} -1 & 2 & 1 & 0 \\ 3 & -3 & -9 & 0 \\ -2 & -1 & 12 & 0 \end{array}\right) \xrightarrow[R_1(-2)+R_3]{R_1(3)+R_2} \left(\begin{array}{ccc|c} -1 & 2 & 1 & 0 \\ 0 & 3 & -6 & 0 \\ 0 & -5 & 10 & 0 \end{array}\right)$

$$\xrightarrow[R_3(-\frac{1}{5})]{R_1(-1),\,R_2(\frac{1}{3})} \left(\begin{array}{ccc|c} 1 & -2 & -1 & 0 \\ 0 & 1 & -2 & 0 \\ 0 & 1 & -2 & 0 \end{array}\right) \xrightarrow{R_2(-1)+R_3}$$

$$\left(\begin{array}{ccc|c} 1 & -2 & -1 & 0 \\ 0 & 1 & -2 & 0 \\ 0 & 0 & 0 & 0 \end{array}\right) \xrightarrow{R_2(2)+R_1} \left(\begin{array}{ccc|c} 1 & 0 & -5 & 0 \\ 0 & 1 & -2 & 0 \\ 0 & 0 & 0 & 0 \end{array}\right).$$

This implies

$$\begin{cases} x_1 - 5x_3 = 0 \\ x_2 - 2x_3 = 0. \end{cases}$$

Let $x_3 = t$. Then $x_1 = 5t$, $x_2 = 2t$, $x_3 = t$ is a solution of th esystem for each $t \in \mathbb{R}$.. □

Section 4.9

1. Solve each of the following systems using the inverse matrix method.

$$a) \begin{cases} x_1 + x_2 + x_3 = 2 \\ 2x_1 + 3x_2 + x_3 = 2 \\ x_1 - x_3 = 1 \end{cases} \qquad b) \begin{cases} x_1 - 2x_2 + 2x_3 = 1 \\ 2x_1 + 3x_2 + 3x_3 = -1 \\ x_1 + 6x_3 = -4 \end{cases}$$

$$c) \begin{cases} x_2 + x_3 = 4 \\ x_1 + x_2 + x_4 = 5 \\ 4x_1 + x_2 + x_3 = 4 \\ -x_1 - 2x_2 + 3x_4 = 0 \end{cases}$$

Solution.

$$a) \quad (A|I) = \left(\begin{array}{ccc|ccc} 1 & 1 & 1 & 1 & 0 & 0 \\ 2 & 3 & 1 & 0 & 1 & 0 \\ 1 & 0 & -1 & 0 & 0 & 1 \end{array}\right) \xrightarrow[R_1(-1)+R_3]{R_1(-2)+R_2}$$

$$\left(\begin{array}{ccc|ccc} 1 & 1 & 1 & 1 & 0 & 0 \\ 0 & 1 & -1 & -2 & 1 & 0 \\ 0 & -1 & -2 & -1 & 0 & 1 \end{array}\right) \xrightarrow{R_2(1)+R_3}$$

$$\left(\begin{array}{ccc|ccc} 1 & 1 & 1 & 1 & 0 & 0 \\ 0 & 1 & -1 & -2 & 1 & 0 \\ 0 & 0 & -3 & -3 & 1 & 1 \end{array}\right) \xrightarrow{R_3(-\frac{1}{3})} \left(\begin{array}{ccc|ccc} 1 & 1 & 1 & 1 & 0 & 0 \\ 0 & 1 & -1 & -2 & 1 & 0 \\ 0 & 0 & 1 & 1 & -\frac{1}{3} & -\frac{1}{3} \end{array}\right)$$

$$\xrightarrow[R_3(1)+R_2]{R_3(-1)+R_1} \left(\begin{array}{ccc|ccc} 1 & 1 & 0 & 0 & \frac{1}{3} & \frac{1}{3} \\ 0 & 1 & 0 & -1 & \frac{2}{3} & -\frac{1}{3} \\ 0 & 0 & 1 & 1 & -\frac{1}{3} & -\frac{1}{3} \end{array}\right)$$

$$\xrightarrow{R_2(-1)+R_1} \left(\begin{array}{ccc|ccc} 1 & 0 & 0 & 1 & -\frac{1}{3} & \frac{2}{3} \\ 0 & 1 & 0 & -1 & \frac{2}{3} & -\frac{1}{3} \\ 0 & 0 & 1 & 1 & -\frac{1}{3} & -\frac{1}{3} \end{array}\right).$$

Hence, $A^{-1} = \begin{pmatrix} 1 & -\frac{1}{3} & \frac{2}{3} \\ -1 & \frac{2}{3} & -\frac{1}{3} \\ 1 & -\frac{1}{3} & -\frac{1}{3} \end{pmatrix}$ and

$$\begin{pmatrix} x_1 \\ x_2 \\ x_3 \end{pmatrix} = \begin{pmatrix} 1 & -\frac{1}{3} & \frac{2}{3} \\ -1 & \frac{2}{3} & -\frac{1}{3} \\ 1 & -\frac{1}{3} & -\frac{1}{3} \end{pmatrix} \begin{pmatrix} 2 \\ 2 \\ 1 \end{pmatrix} = \begin{pmatrix} 2 - \frac{2}{3} + \frac{2}{3} \\ -2 + \frac{4}{3} - \frac{1}{3} \\ 2 - \frac{2}{3} - \frac{1}{3} \end{pmatrix} = \begin{pmatrix} 2 \\ -1 \\ 1 \end{pmatrix}.$$

b) $$(A|I) = \left(\begin{array}{ccc|ccc} 1 & -2 & 2 & 1 & 0 & 0 \\ 2 & 3 & 3 & 0 & 1 & 0 \\ 1 & 0 & 6 & 0 & 0 & 1 \end{array}\right) \xrightarrow[R_1(-1)+R_3]{R_1(-2)+R_2}$$

$$\left(\begin{array}{ccc|ccc} 1 & -2 & 2 & 1 & 0 & 0 \\ 0 & 7 & -1 & -2 & 1 & 0 \\ 0 & 2 & 4 & -1 & 0 & 1 \end{array}\right) \xrightarrow{R_3(-3)+R_2}$$

$$\left(\begin{array}{ccc|ccc} 1 & -2 & 2 & 1 & 0 & 0 \\ 0 & 1 & -13 & 1 & 1 & -3 \\ 0 & 2 & 4 & -1 & 0 & 1 \end{array}\right) \xrightarrow{R_2(-2)+R_3}$$

$$\left(\begin{array}{ccc|ccc} 1 & -2 & 2 & 1 & 0 & 0 \\ 0 & 1 & -13 & 1 & 1 & -3 \\ 0 & 0 & 30 & -3 & -2 & 7 \end{array}\right) \xrightarrow[R_3(\frac{1}{30})]{R_2(2)+R_1}$$

$$\left(\begin{array}{ccc|ccc} 1 & 0 & -24 & 3 & 2 & -6 \\ 0 & 1 & -13 & 1 & 1 & -3 \\ 0 & 0 & 1 & -\frac{1}{10} & -\frac{1}{15} & \frac{7}{30} \end{array}\right) \xrightarrow[R_3(13)+R_2]{R_3(24)+R_1}$$

$$\left(\begin{array}{ccc|ccc} 1 & 0 & 0 & 3-\frac{24}{10} & 2-\frac{24}{15} & -6+24\frac{7}{30} \\ 0 & 1 & 0 & 1-\frac{13}{10} & 1-\frac{13}{15} & -3+13\frac{7}{30} \\ 0 & 0 & 1 & -\frac{1}{10} & -\frac{1}{15} & \frac{7}{30} \end{array}\right)$$

$$= \left(\begin{array}{ccc|ccc} 1 & 0 & 0 & \frac{3}{5} & \frac{2}{5} & -\frac{2}{5} \\ 0 & 1 & 0 & -\frac{3}{10} & \frac{2}{15} & \frac{1}{30} \\ 0 & 0 & 1 & -\frac{1}{10} & -\frac{1}{15} & \frac{7}{30} \end{array}\right).$$

Hence, $A^{-1} = \begin{pmatrix} \frac{3}{5} & \frac{2}{5} & -\frac{2}{5} \\ -\frac{3}{10} & \frac{2}{15} & \frac{1}{30} \\ -\frac{1}{10} & -\frac{1}{15} & \frac{7}{30} \end{pmatrix}$ and

$$\begin{pmatrix} x_1 \\ x_2 \\ x_3 \end{pmatrix} = A^{-1}\overrightarrow{b} = \begin{pmatrix} \frac{3}{5} & \frac{2}{5} & -\frac{2}{5} \\ -\frac{3}{10} & \frac{2}{15} & \frac{1}{30} \\ -\frac{1}{10} & -\frac{1}{15} & \frac{7}{30} \end{pmatrix} \begin{pmatrix} 1 \\ -1 \\ -4 \end{pmatrix} = \begin{pmatrix} \frac{3}{5} - \frac{2}{5} + \frac{8}{5} \\ -\frac{3}{10} - \frac{2}{15} - \frac{4}{30} \\ -\frac{1}{10} + \frac{1}{15} - \frac{28}{30} \end{pmatrix}$$

$$= \begin{pmatrix} \frac{9}{5} \\ -\frac{17}{30} \\ -\frac{29}{30} \end{pmatrix}.$$

c) $(A|I) = \left(\begin{array}{cccc|cccc} 0 & 1 & 1 & 0 & 1 & 0 & 0 & 0 \\ 1 & 1 & 0 & 1 & 0 & 1 & 0 & 0 \\ 4 & 1 & 1 & 0 & 0 & 0 & 1 & 0 \\ -1 & -2 & 0 & 3 & 0 & 0 & 0 & 1 \end{array}\right) \xrightarrow{R_{1,2}}$

$$\left(\begin{array}{cccc|cccc} 1 & 1 & 0 & 1 & 0 & 1 & 0 & 0 \\ 0 & 1 & 1 & 0 & 1 & 0 & 0 & 0 \\ 4 & 1 & 1 & 0 & 0 & 0 & 1 & 0 \\ -1 & -2 & 0 & 3 & 0 & 0 & 0 & 1 \end{array}\right) \xrightarrow[R_1(1)+R_4]{R_1(-4)+R_3}$$

$$\left(\begin{array}{cccc|cccc} 1 & 1 & 0 & 1 & 0 & 1 & 0 & 0 \\ 0 & 1 & 1 & 0 & 1 & 0 & 0 & 0 \\ 0 & -3 & 1 & -4 & 0 & -4 & 1 & 0 \\ 0 & -1 & 3 & 1 & 0 & 1 & 0 & 1 \end{array}\right) \xrightarrow[R_2(1)+R_4]{R_2(3)+R_3}$$

$$\left(\begin{array}{cccc|cccc} 1 & 1 & 0 & 1 & 0 & 1 & 0 & 0 \\ 0 & 1 & 1 & 0 & 1 & 0 & 0 & 0 \\ 0 & 0 & 4 & -4 & 3 & -4 & 1 & 0 \\ 0 & 0 & 4 & 1 & 1 & 1 & 0 & 1 \end{array}\right) \xrightarrow{R_3(-1)+R_4}$$

$$\left(\begin{array}{cccc|cccc} 1 & 1 & 0 & 1 & 0 & 1 & 0 & 0 \\ 0 & 1 & 1 & 0 & 1 & 0 & 0 & 0 \\ 0 & 0 & 4 & -4 & 3 & -4 & 1 & 0 \\ 0 & 0 & 0 & 5 & -2 & 5 & -1 & 1 \end{array}\right) \xrightarrow{R_4(\frac{1}{5})}$$

$$\left(\begin{array}{cccc|cccc} 1 & 1 & 0 & 1 & 0 & 1 & 0 & 0 \\ 0 & 1 & 1 & 0 & 1 & 0 & 0 & 0 \\ 0 & 0 & 4 & -4 & 3 & -4 & 1 & 0 \\ 0 & 0 & 0 & 1 & -\frac{2}{5} & 1 & -\frac{1}{5} & \frac{1}{5} \end{array}\right) \xrightarrow[R_4(4)+R_3]{R_4(-1)+R_1}$$

$$\left(\begin{array}{cccc|cccc} 1 & 1 & 0 & 0 & \frac{2}{5} & 0 & \frac{1}{5} & -\frac{1}{5} \\ 0 & 1 & 1 & 0 & 1 & 0 & 0 & 0 \\ 0 & 0 & 4 & 0 & \frac{7}{5} & 0 & \frac{1}{5} & \frac{4}{5} \\ 0 & 0 & 0 & 1 & -\frac{2}{5} & 1 & -\frac{1}{5} & \frac{1}{5} \end{array}\right) \xrightarrow{R_3(\frac{1}{4})}$$

$$\left(\begin{array}{cccc|cccc} 1 & 1 & 0 & 0 & \frac{2}{5} & 0 & \frac{1}{5} & -\frac{1}{5} \\ 0 & 1 & 1 & 0 & 1 & 0 & 0 & 0 \\ 0 & 0 & 1 & 0 & \frac{7}{20} & 0 & \frac{1}{20} & \frac{1}{5} \\ 0 & 0 & 0 & 1 & -\frac{2}{5} & 1 & -\frac{1}{5} & \frac{1}{5} \end{array}\right) \xrightarrow{R_3(-1)+R_2}$$

$$\left(\begin{array}{cccc|cccc} 1 & 1 & 0 & 0 & \frac{2}{5} & 0 & \frac{1}{5} & -\frac{1}{5} \\ 0 & 1 & 0 & 0 & \frac{13}{20} & 0 & -\frac{1}{20} & -\frac{1}{5} \\ 0 & 0 & 1 & 0 & \frac{7}{20} & 0 & \frac{1}{20} & \frac{1}{5} \\ 0 & 0 & 0 & 1 & -\frac{2}{5} & 1 & -\frac{1}{5} & \frac{1}{5} \end{array}\right) \xrightarrow{R_2(-1)+R_1}$$

$$\left(\begin{array}{cccc|cccc} 1 & 0 & 0 & 0 & -\frac{1}{4} & 0 & \frac{1}{4} & 0 \\ 0 & 1 & 0 & 0 & \frac{13}{20} & 0 & -\frac{1}{20} & -\frac{1}{5} \\ 0 & 0 & 1 & 0 & \frac{7}{20} & 0 & \frac{1}{20} & \frac{1}{5} \\ 0 & 0 & 0 & 1 & -\frac{2}{5} & 1 & -\frac{1}{5} & \frac{1}{5} \end{array}\right).$$

Hence, $A^{-1} = \begin{pmatrix} -\frac{1}{4} & 0 & \frac{1}{4} & 0 \\ \frac{13}{20} & 0 & -\frac{1}{20} & -\frac{1}{5} \\ \frac{7}{20} & 0 & \frac{1}{20} & \frac{1}{5} \\ -\frac{2}{5} & 1 & -\frac{1}{5} & \frac{1}{5} \end{pmatrix}$ and

$$\begin{pmatrix} x_1 \\ x_2 \\ x_3 \\ x_4 \end{pmatrix} = A^{-1}\vec{b} = \begin{pmatrix} -\frac{1}{4} & 0 & \frac{1}{4} & 0 \\ \frac{13}{20} & 0 & -\frac{1}{20} & -\frac{1}{5} \\ \frac{7}{20} & 0 & \frac{1}{20} & \frac{1}{5} \\ -\frac{2}{5} & 1 & -\frac{1}{5} & \frac{1}{5} \end{pmatrix} \begin{pmatrix} 4 \\ 5 \\ 4 \\ 0 \end{pmatrix} = \begin{pmatrix} 0 \\ \frac{12}{5} \\ \frac{8}{5} \\ \frac{13}{5} \end{pmatrix}.$$

□

Section 4.10

1. Solve the following system by using Cramer's rule.

$a) \begin{cases} x - 3y = 6 \\ 2x + 5y = -1. \end{cases}$ $\quad b) \begin{cases} x_1 + 2x_3 = 2 \\ -2x_1 + 3x_2 + x_3 = 16 \\ -x_1 - 2x_2 + 4x_3 = 5. \end{cases}$

$c) \begin{cases} x_1 + x_2 - x_3 = 1 \\ 2x_1 + x_2 - x_3 = 0 \\ x_2 - 2x_3 = -2. \end{cases}$

Solution.

a). $|A| = \begin{vmatrix} 1 & -3 \\ 2 & 5 \end{vmatrix} = 5 - (-6) = 11$, $|A_1| = \begin{vmatrix} 6 & -3 \\ -1 & 5 \end{vmatrix} = 30 - 3 = 27$ and $|A_2| = \begin{vmatrix} 1 & 6 \\ 2 & -1 \end{vmatrix} = -1 - 12 = -13$. Hence,

$$x = \frac{|A_1|}{|A|} = \frac{27}{11} \quad \text{and } y = \frac{|A_2|}{|A|} = -\frac{13}{11}.$$

b). $|A| = \left|\begin{matrix} 1 & 0 & 2 \\ -2 & 3 & 1 \\ -1 & -2 & 4 \end{matrix}\right| \begin{matrix} 1 & 0 \\ -2 & 3 \\ -1 & -2 \end{matrix} = (12 + 8) - (-6 - 2) = 28.$

$|A_1| = \left|\begin{matrix} 2 & 0 & 2 \\ 16 & 3 & 1 \\ 5 & -2 & 4 \end{matrix}\right| \begin{matrix} 2 & 0 \\ 16 & 3 \\ 5 & -2 \end{matrix} = (24 - 64) - (30 - 4) = 66.$

$|A_2| = \left|\begin{matrix} 1 & 2 & 2 \\ -2 & 16 & 1 \\ -1 & 5 & 4 \end{matrix}\right| \begin{matrix} 1 & 2 \\ -2 & 16 \\ -1 & 5 \end{matrix} = (64 - 2 - 20) - (-32 + 5 - 16) = 85.$

$$|A_3| = \begin{vmatrix} 1 & 0 & 2 \\ -2 & 3 & 16 \\ -1 & -2 & 5 \end{vmatrix} \begin{matrix} 1 & 0 \\ -2 & 3 \\ -1 & -2 \end{matrix} = (15+8) - (-6-32) = 61. \text{ Hence,}$$

$$x_1 = \frac{|A_1|}{|A|} = \frac{66}{28} = \frac{33}{14}, \quad x_2 = \frac{|A_2|}{|A|} = \frac{85}{28} \text{ and } x_3 = \frac{|A_3|}{|A|} = \frac{61}{28}.$$

c). $|A| = \begin{vmatrix} 1 & 1 & -1 \\ 2 & 1 & -1 \\ 0 & 1 & -2 \end{vmatrix} \begin{matrix} 1 & 1 \\ 2 & 1 \\ 0 & 1 \end{matrix} = (-2-2) - (-1-4) = 1.$

$$|A_1| = \begin{vmatrix} 1 & 1 & -1 \\ 0 & 1 & -1 \\ -2 & 1 & -2 \end{vmatrix} \begin{matrix} 1 & 1 \\ 0 & 1 \\ -2 & 1 \end{matrix} = (-2+2) - (2-1) = -1.$$

$$|A_2| = \begin{vmatrix} 1 & 1 & -1 \\ 2 & 0 & -1 \\ 0 & -2 & -2 \end{vmatrix} \begin{matrix} 1 & 1 \\ 2 & 0 \\ 0 & -2 \end{matrix} = 4 - (2-4) = 6.$$

$$|A_3| = \begin{vmatrix} 1 & 1 & 1 \\ 2 & 1 & 0 \\ 0 & 1 & -2 \end{vmatrix} \begin{matrix} 1 & 1 \\ 2 & 1 \\ 0 & 1 \end{matrix} = (-2+2) - (-4) = 4. \text{ Hence}$$

$$x_1 = \frac{|A_1|}{|A|} = \frac{-1}{1} = -1, \quad x_2 = \frac{|A_2|}{|A|} = \frac{6}{1} = 6 \text{ and } x_3 = \frac{|A_3|}{|A|} = \frac{4}{1} = 4.$$

□

A.5 vectors in $\mathbb{R}^2$ and $\mathbb{R}^2$

Section 5.1

1. Find $\|\overrightarrow{u}\|$.

   (1) $\overrightarrow{u} = (1, 0, -2)$; (2) $\overrightarrow{u} = (1, 1, -2)$; (3) $\overrightarrow{u} = (2, 2, -2)$.

2. Find the distance between P_1 and P_2.

   (1) $P_1(1, -1, 5)$ and $P_2(1, -3, 1)$; (2) $P_1(0, -1, 2)$ and $P_2(1, -3, 1)$

3. Find the cosine of the angle θ between $\overrightarrow{u}$ and $\overrightarrow{v}$.

   (1) $\overrightarrow{u} = (1, -1, 1)$ and $\overrightarrow{v} = (1, 1, -2)$;

   (2) $\overrightarrow{u} = (1, 0, 1)$ and $\overrightarrow{v} = (-1, -1, -1)$.

4. Find $\text{proj}_{\overrightarrow{a}} \overrightarrow{u}$.

   (1) $\overrightarrow{u} = (1, -1, 2)$ and $\overrightarrow{a} = (2, -1, 1)$;

   (2) $\overrightarrow{u} = (1, 0, -1)$ and $\overrightarrow{a} = (-1, -2, 2)$.

Solution. 1. (1) $\|\overrightarrow{u}\| = \sqrt{1^2+0^2+(-2)^2} = \sqrt{5}$.
(2) $\|\overrightarrow{u}\| = \sqrt{1^2+1^2+(-2)^2} = \sqrt{6}$.
(3) $\|\overrightarrow{u}\| = \sqrt{2^2+2^2+(-2)^2} = \sqrt{12} = 2\sqrt{3}$.

2. (1) $d(P_1P2) = \|\overrightarrow{P_1P_2}\| = \|(1-1,-3+1,1-5)\| = \|(0,-2,-4)\|$
$= \sqrt{0^2+(-2)^2+(-4)^2} = 2\sqrt{5}$.

(2) $d(P_1P2) = \|\overrightarrow{P_1P_2}\| = \|(1-0,-3+1,1-2)\| = \|(1,-2,-1)\|$
$= \sqrt{1^2+(-2)^2+(-1)^2} = \sqrt{6}$.

3. (1) Since $\overrightarrow{u}\cdot\overrightarrow{v} = (1)(1)+(-1)(1)+(1)(-2) = -2$,

$$\|\overrightarrow{u}\| = \sqrt{1^2+(-1)^2+1^2} = \sqrt{3} \quad \text{and } \|\overrightarrow{v}\| = \sqrt{1^2+(1)^2+(-2)^2} = \sqrt{6},$$

by Theorem 5.1.3, we have

$$\cos\theta = \frac{\overrightarrow{u}\cdot\overrightarrow{v}}{\|\overrightarrow{u}\|\|\overrightarrow{v}\|} = \frac{-2}{\sqrt{3}\sqrt{6}} = -\frac{2}{3\sqrt{2}} = -\frac{\sqrt{2}}{3}.$$

(2) Since $\overrightarrow{u}\cdot\overrightarrow{v} = (1)(-1)+(0)(-1)+(1)(-1) = -2$,

$$\|\overrightarrow{u}\| = \sqrt{1^2+(0)^2+1^2} = \sqrt{2} \quad \text{and } \|\overrightarrow{v}\| = \sqrt{(-1)^2+(-1)^2+(-1)^2} = \sqrt{3},$$

by Theorem 5.1.3, we have

$$\cos\theta = \frac{\overrightarrow{u}\cdot\overrightarrow{v}}{\|\overrightarrow{u}\|\|\overrightarrow{v}\|} = \frac{-2}{\sqrt{2}\sqrt{3}} = -\frac{2}{\sqrt{6}} = -\frac{\sqrt{6}}{3}.$$

4. (1) $\text{proj}_{\overrightarrow{a}}\overrightarrow{u} = (\frac{\overrightarrow{u}\cdot\overrightarrow{a}}{\|\overrightarrow{a}\|^2})\overrightarrow{a} = \frac{(1)(2)+(-1)(-1)+(2)(1)}{2^2+(-1)^2+1^2}(2,-1,1)$
$= \frac{5}{6}(2,-1,1) = (\frac{5}{3},-\frac{5}{6},\frac{5}{6})$.

(2) $\text{proj}_{\overrightarrow{a}}\overrightarrow{u} = (\frac{\overrightarrow{u}\cdot\overrightarrow{a}}{\|\overrightarrow{a}\|^2})\overrightarrow{a} = \frac{(1)(-1)+(0)(-2)+(-1)(2)}{(-1)^2+(-2)^2+2^2}(-1,-2,2)$
$= \frac{-3}{9}(-1,-2,2) = -(-1,-2,2) = (1,2,-2)$.

□

Section 5.2

1. Find the cross product $\overrightarrow{u}\times\overrightarrow{v}$.

(1) $\overrightarrow{u} = (1,2,3)$; $\overrightarrow{v} = (1,0,1)$, (2) $\overrightarrow{u} = (1,0,-3)$; $\overrightarrow{v} = (-1,0,-2)$
(3) $\overrightarrow{u} = (0,2,1)$; $\overrightarrow{v} = (-5,0,1)$, (4) $\overrightarrow{u} = (1,1,-1)$; $\overrightarrow{v} = (-1,1,0)$.

2. Find the area of the parallelogram determined by $\overrightarrow{u}$ and $\overrightarrow{v}$.

   (1) $\overrightarrow{u} = (1, 0, 1)$; $\overrightarrow{v} = (-1, 0, 1)$, (2) $\overrightarrow{u} = (1, 0, 0)$; $\overrightarrow{v} = (-1, 1, -2)$
   (3) $\overrightarrow{u} = (0, 2, -1)$; $\overrightarrow{v} = (-5, 1, 1)$, (4) $\overrightarrow{u} = (1, -1, -1)$; $\overrightarrow{v} = (-1, 1, 2)$.

3. Find the area of the parallelogram determined by $\overrightarrow{u}$ and $\overrightarrow{v}$.

   (1) $\overrightarrow{u} = (1, 0)$; $\overrightarrow{v} = (0, -1)$, (2) $\overrightarrow{u} = (1, 0)$; $\overrightarrow{v} = (-1, 1)$
   (3) $\overrightarrow{u} = (0, 2)$; $\overrightarrow{v} = (1, 1)$, (4) $\overrightarrow{u} = (1, -1)$; $\overrightarrow{v} = (-1, 2)$.

4. Find the volume of the parallelpiped determined by $\overrightarrow{u}$, $\overrightarrow{u}$ and $\overrightarrow{u}$.

   (1) $\overrightarrow{u} = (1, -1, 0)$, $\overrightarrow{v} = (-1, 0, 2)$ and $\overrightarrow{w} = (0, -1, -1)$.

   (2) $\overrightarrow{u} = (-1, -1, 0)$, $\overrightarrow{v} = (-1, 1, -2)$ and $\overrightarrow{w} = (-1, 1, 1)$.

Solution. 1 (1) $\overrightarrow{u} \times \overrightarrow{v} = \left(\begin{vmatrix} 2 & 3 \\ 0 & 1 \end{vmatrix}, - \begin{vmatrix} 1 & 3 \\ 1 & 1 \end{vmatrix}, \begin{vmatrix} 1 & 2 \\ 1 & 0 \end{vmatrix} \right) = (2, 2, -2)$.

(2) $\overrightarrow{u} \times \overrightarrow{v} = \left(\begin{vmatrix} 0 & -3 \\ 0 & -2 \end{vmatrix}, - \begin{vmatrix} 1 & -3 \\ -1 & -2 \end{vmatrix}, \begin{vmatrix} 1 & 0 \\ -1 & 0 \end{vmatrix} \right) = (0, 5, 0)$.

(3) $\overrightarrow{u} \times \overrightarrow{v} = \left(\begin{vmatrix} 2 & 1 \\ 0 & 1 \end{vmatrix}, - \begin{vmatrix} 0 & 1 \\ -5 & 1 \end{vmatrix}, \begin{vmatrix} 0 & 2 \\ -5 & 0 \end{vmatrix} \right) = (2, 5, -10)$.

(4) $\overrightarrow{u} \times \overrightarrow{v} = \left(\begin{vmatrix} 1 & -1 \\ 1 & 0 \end{vmatrix}, - \begin{vmatrix} 1 & -1 \\ -1 & 0 \end{vmatrix}, \begin{vmatrix} 1 & 1 \\ -1 & 1 \end{vmatrix} \right) = (-1, -1, 2)$.

2 (1) $\overrightarrow{u} \times \overrightarrow{v} = \left(\begin{vmatrix} 0 & 1 \\ 0 & 1 \end{vmatrix}, - \begin{vmatrix} 1 & 1 \\ -1 & 1 \end{vmatrix}, \begin{vmatrix} 1 & 0 \\ -1 & 0 \end{vmatrix} \right) = (0, -2, 0)$. Hence,

$$A = \sqrt{0^2 + (-2)^2 + 0^2} = 2.$$

(2) $\overrightarrow{u} \times \overrightarrow{v} = \left(\begin{vmatrix} 0 & 0 \\ 1 & -2 \end{vmatrix}, - \begin{vmatrix} 1 & 0 \\ -1 & -2 \end{vmatrix}, \begin{vmatrix} 1 & 0 \\ -1 & 1 \end{vmatrix} \right) = (0, 2, 1)$. Hence,

$$A = \sqrt{0^2 + 2^2 + 1^2} = \sqrt{5}.$$

(3) $\overrightarrow{u} \times \overrightarrow{v} = \left(\begin{vmatrix} 2 & -1 \\ 1 & 1 \end{vmatrix}, - \begin{vmatrix} 0 & -1 \\ -5 & 1 \end{vmatrix}, \begin{vmatrix} 0 & 2 \\ -5 & 1 \end{vmatrix} \right) = (3, -5, 10)$. Hence,

$$A = \sqrt{3^2 + (-5)^2 + 10^2} = \sqrt{136} = 2\sqrt{34}.$$

(4) $\overrightarrow{u} \times \overrightarrow{v} = \left(\begin{vmatrix} -1 & -1 \\ 1 & 2 \end{vmatrix}, - \begin{vmatrix} 1 & -1 \\ -1 & 2 \end{vmatrix}, \begin{vmatrix} 1 & -1 \\ -1 & 1 \end{vmatrix} \right) = (-1, -1, 0)$.
Hence,

$$A = \sqrt{(-1)^2 + (-1)^2 + 0^2} = \sqrt{2}.$$

3 (1) Since $\begin{vmatrix} 1 & 0 \\ 0 & -1 \end{vmatrix} = -1$, $A = |-1| = 1$.

(2) Since $\begin{vmatrix} 1 & 0 \\ -1 & 1 \end{vmatrix} = 1$, $A = |1| = 1$.

(3) Since $\begin{vmatrix} 0 & 2 \\ 1 & 1 \end{vmatrix} = -2$, $A = |-2| = 2$.

(1) Since $\begin{vmatrix} 1 & -1 \\ -1 & 2 \end{vmatrix} = 1$, $A = |1| = 1$.

4 (1) Since $\begin{vmatrix} 1 & -1 & 0 \\ -1 & 0 & 2 \\ 0 & -1 & -1 \end{vmatrix} = 2$, $V = |2| = 2$.

(2) Since $\begin{vmatrix} -1 & -1 & 0 \\ -1 & 1 & -2 \\ -1 & 1 & 1 \end{vmatrix} = -6$, $V = |-6| = 6$.

□

Section 5.3

1. Find an equation for the plane through $P_0(1,1,3)$ that is perpendicular to the line

$$\begin{cases} x = 2 - 3t \\ y = 1 + t \\ z = 2t \end{cases}$$

2. Find an equation for the plane through $P_0(2,7,-1)$ that is parallel to the plane $4x - y + 3z = 3$.

3. Find and equation for the plane that contains the line $x = 3 + t$, $y = 5$, $z = t + 2t$, and is perpendicular to the plane $x + y + z = 4$.

4. Find an equation for the plane through $P_0(1,4,4)$ that contains the line of intersection of the planes

$$\begin{cases} x - y + 3z = 5 \\ 2x + 2y + 7z = 0 \end{cases}$$

5. Find an equation for the plane containing the line $x = 3 + 6t$, $y = 4$, $z = t$ and that is parallel to the line of intersection of the plane $2x + y + z = 1$ and $x - 2y + 3z = 2$.

Solution. 1. $\vec{n} = \vec{v} = (-3,1,2)$, so $-3(x-1) + (y-1) + 2(z-3) = 0$ or $-3x + y + 2z = 4$.

2. Since $P_0(2,7,-1)$ and $\vec{n} = (4,-1,3)$, the equation is $4x - y + 3z = -2$.

3 Since $P_0(3,5,5)$ and $\vec{v} = (1,0,2)$,

$$\vec{n} = \vec{v} \times \vec{n_1} = (1,0,2) \times (1,1,1) = (-2,1,1)$$

and $-2(x-3)+(y-5)+(z-5)=0$ or $-2x+y+z=4$.

4 We find $\vec{n}$. Solving (1) and (2),

$$\begin{cases} x = \frac{5}{2} - \frac{13}{4}t \\ y = -\frac{5}{2} - \frac{1}{4}t \\ z = t \end{cases} \tag{A.2}$$

we get $\vec{v} = (-\frac{13}{4}, -\frac{1}{4}, 1)$ and $P_1 = (\frac{5}{2}, -\frac{5}{2}, 0)$. This, together with $P_0(1,4,4)$, implies

$$\vec{n} = \vec{v} \times \overrightarrow{P_0P_1} = (-\frac{13}{4}, -\frac{1}{4}, 1) \times (\frac{3}{2}, -\frac{13}{2}, -4) = \frac{1}{2}(15, -23, 43).$$

Hence, $15(x-1) - 23(y-4) + 43(z-4) = 0$ or $15x - 23y + 43z = 95$.

5. $P_0 = (3,4,0)$, $\vec{v_1} = (16,0,1)$, $\vec{v_2} = (2,1,1) \times (1,-2,3) = (1,1,1)$.

$\vec{n} = \vec{v_2} \times \vec{v_1} = (-1,-7,6)$. Hence, $-1(x-3) - 7(y-4) + 6(z-0) = 0$ or $x + 7y - 6z = 31$. □

Index